Stochastic Empirical Loading and Dilution Model (SELDM) Version 1.0.0

By Gregory E. Granato

Chapter 3
Section C, Water Quality,
Book 4, Hydrologic Analysis and Interpretation

Prepared in cooperation with the
U.S. Department of Transportation, Federal Highway Administration,
Office of Project Development and Environmental Review

Techniques and Methods 4–C3

U.S. Department of the Interior
U.S. Geological Survey

U.S. Department of the Interior
KEN SALAZAR, Secretary

U.S. Geological Survey
Suzette M. Kimball, Acting Director

U.S. Geological Survey, Reston, Virginia: 2013

For more information on the USGS—the Federal source for science about the Earth, its natural and living resources, natural hazards, and the environment, visit http://www.usgs.gov or call 1–888–ASK–USGS.

For an overview of USGS information products, including maps, imagery, and publications, visit http://www.usgs.gov/pubprod

To order this and other USGS information products, visit http://store.usgs.gov

Suggested citation:
Granato, G.E., 2013, Stochastic empirical loading and dilution model (SELDM) version 1.0.0: U.S. Geological Survey Techniques and Methods, book 4, chap. C3, 112 p., CD–ROM. (Also available at http://pubs.usgs.gov/tm/04/c03/.)

Acknowledgments

The author thanks the many people who reviewed this manual, the model, and the associated CD–ROM. Susan Jones of the Federal Highway Administration (FHWA), John Risley, Robert Runkel, Mary Ashman, and Kevin Breen of the U.S. Geological Survey (USGS), David Graves of the New York State Department of Transportation (DOT), William Fletcher of the Oregon DOT, and Rachel Herbert of the U.S. Environmental Protection Agency (USEPA) for providing thoughtful and thorough technical and editorial reviews of this report, the software, appendixes, and the associated CD–ROM. Elizabeth Ahearn and Anthony Gotvald of the USGS and Joseph Krolak of the FHWA reviewed the information supporting development and use of the basin-lagtime equations and the triangular hydrographs. Richard Vogel of Tufts University provided information on use of the probability-plot correlation coefficient method for evaluating the distribution of the random numbers generated by SELDM.

The author also thanks the many people who helped ensure that the model and the graphical user interface work correctly. William Fletcher of the Oregon DOT, Mark Mattson of the Massachusetts Department of Environmental Protection, Andrew McDaniel of the North Carolina DOT, and David Graves of the New York DOT provided exceptionally detailed reviews for several beta-test versions of the model. Patricia Cazenas, Susan Jones, Cynthia Nurmi, and Brian Smith of the FHWA participated in the beta tests. Rachel Herbert, Dino Marshalonis, Jesse Pritts, Nelly Smith, and Mark Sievers of the USEPA; Ryan McReynolds of the U.S. Fish and Wildlife Service; and Jeffrey Chaplin, Judy Horwatich, and John Risley of the USGS also participated in one or more beta tests of the model. Additionally, environmental professionals from 14 other State DOTs participated in one or more beta tests: Kristine Benson and Guangyan Griffin from Alaska; Dennis Cress, Rik Gay, and Holly Huyck from Colorado; Steven Sisson from Delaware; Karen Coffman and Ling Li from Maryland; Henry Barbaro, Ying Jiang, and Alex Murray from Massachusetts; John Taylor from Mississippi; James Murphy from Nevada; Paul Kieda from New York; Matthew Lauffer from North Carolina; Michele Dolan from Oklahoma; Crystal Newcomer from Pennsylvania; Allison Drake and Allison Hamel from Rhode Island; Alex Nguyen, Le Nguyen, and Kristianne Sandoval from Washington; and Jackie Fields from West Virginia.

Contents

Figures

Tables

Conversion Factors, Datum, and Abbreviations

Inch/Pound to SI

Multiply	By	To obtain
Length		
inch (in.)	2.54	centimeter (cm)
foot (ft)	0.3048	meter (m)
mile (mi)	1.609	kilometer (km)
Area		
acre	4,047	square meter (m^2)
acre	0.4047	hectare (ha)
square foot (ft^2)	0.09290	square meter (m^2)
square inch (in^2)	6.452	square centimeter (cm^2)
square mile (mi^2)	259.0	hectare (ha)
square mile (mi^2)	2.590	square kilometer (km^2)
Volume		
cubic foot (ft^3)	28.32	liter (L)
cubic foot (ft^3)	0.02832	cubic meter (m^3)
Flow rate		
foot per year (ft/yr)	0.3048	meter per year (m/yr)
cubic foot per second (ft^3/s)	0.02832	cubic meter per second (m^3/s)
cubic foot per second per square mile [(ft^3/s)/mi^2]	0.01093	cubic meter per second per square kilometer [(m^3/s)/km^2]
inch per hour (in/h)	0.0254	meter per hour (m/h)
Mass		
pound, avoirdupois (lb)	0.4536	kilogram (kg)
Density		
pound per cubic foot (lb/ft^3)	0.01602	gram per cubic centimeter (g/cm^3)

Vertical coordinate information is referenced to the North American Vertical Datum of 1988 (NAVD 88).

Horizontal coordinate information is referenced to the North American Datum of 1983 (NAD 83).

Altitude, as used in this report, refers to distance above the vertical datum.

Temperature in degrees Celsius (°C) may be converted to degrees Fahrenheit (°F) as follows:

$$°F=(1.8×°C)+32$$

Temperature in degrees Fahrenheit (°F) may be converted to degrees Celsius (°C) as follows:

$$°C=(°F–32)/1.8$$

Chemical concentrations in water are given in units of milligrams per liter (mg/L) or micrograms per liter (µg/L), which express the mass of solute per unit volume (liter) of water. Milligrams per liter are equivalent to "parts per million." Micrograms per liter are equivalent to "parts per billion."

Densities of trace elements in sediment are given in units of micrograms per kilogram (µg/kg).

For water-quality loads, 28.32 liters per second (L/s) = 1 cubic foot per second (ft^3/s), and 10.93 liters per second per square kilometer (L/s/km^2) = 1 cubic foot per second per square mile (ft^3/s/mi^2).

Sediment-grain diameters and the diameters of spheres are given in microns.

Scientific notation is used on some graph figures for very large and very small numbers. For example, $1 × 10^{-3}$ and 1E-3 are equivalent to 0.001.

Gravitational acceleration is given in centimeters per second per second (cm/s^2).

Abbreviations

BDF	basin development factor
BLF	basin lag factor
BMP	best management practice
CMRRNG	combined multiple recursive random number generator
CDF	cumulative distribution function
CMC	criteria maximum concentration
COV	coefficient of variation
DOT	department of transportation
DQO	data-quality objectives
EMC	event mean concentration
EPM	effluent probability method
FAV	final acute value
FHWA	Federal Highway Administration
GIS	geographic information system
GUI	graphical user interface
HRDB	Highway Runoff Database
IET	interevent time

IF	impervious fraction
IQR	interquartile range
MAD	median absolute deviation
MPV	most probable value
NPDES	National Pollution Discharge Elimination System
NSQD	National Stormwater Quality Database
NURP	Nationwide Urban Runoff Program
NWISWeb	National Water-Quality Information System Web
NWISSC	National Water Information System Site Cleaner
NWiz	National Water Information System Wizard
PCode	U.S. Environmental Protection Agency parameter code
PDF	probability-density function
PPCC	probability-plot correlation coefficient
RDBP	Relational DataBase File Processor
SSC	suspended-sediment concentration
SELDM	Stochastic Empirical Loading and Dilution Model
STORET	USEPA STOrage and RETrieval database
SWQDM	Surface-Water-Quality Data Miner
TIA	total impervious area
TMDL	Total Maximum Daily Load
TSS	total suspended solids
USEPA	U.S. Environmental Protection Agency
VBA	Visual Basic for Applications®

Stochastic Empirical Loading and Dilution Model (SELDM) Version 1.0.0

By Gregory E. Granato

Abstract

The Stochastic Empirical Loading and Dilution Model (SELDM) is designed to transform complex scientific data into meaningful information about the risk of adverse effects of runoff on receiving waters, the potential need for mitigation measures, and the potential effectiveness of such management measures for reducing these risks. The U.S. Geological Survey developed SELDM in cooperation with the Federal Highway Administration to help develop planning-level estimates of event mean concentrations, flows, and loads in stormwater from a site of interest and from an upstream basin. Planning-level estimates are defined as the results of analyses used to evaluate alternative management measures; planning-level estimates are recognized to include substantial uncertainties (commonly orders of magnitude). SELDM uses information about a highway site, the associated receiving-water basin, precipitation events, stormflow, water quality, and the performance of mitigation measures to produce a stochastic population of runoff-quality variables. SELDM provides input statistics for precipitation, prestorm flow, runoff coefficients, and concentrations of selected water-quality constituents from National datasets. Input statistics may be selected on the basis of the latitude, longitude, and physical characteristics of the site of interest and the upstream basin. The user also may derive and input statistics for each variable that are specific to a given site of interest or a given area.

SELDM is a stochastic model because it uses Monte Carlo methods to produce the random combinations of input variable values needed to generate the stochastic population of values for each component variable. SELDM calculates the dilution of runoff in the receiving waters and the resulting downstream event mean concentrations and annual average lake concentrations. Results are ranked, and plotting positions are calculated, to indicate the level of risk of adverse effects caused by runoff concentrations, flows, and loads on receiving waters by storm and by year. Unlike deterministic hydrologic models, SELDM is not calibrated by changing values of input variables to match a historical record of values. Instead, input values for SELDM are based on site characteristics and representative statistics for each hydrologic variable. Thus, SELDM is an empirical model based on data and statistics rather than theoretical physiochemical equations.

SELDM is a lumped parameter model because the highway site, the upstream basin, and the lake basin each are represented as a single homogeneous unit. Each of these source areas is represented by average basin properties, and results from SELDM are calculated as point estimates for the site of interest. Use of the lumped parameter approach facilitates rapid specification of model parameters to develop planning-level estimates with available data. The approach allows for parsimony in the required inputs to and outputs from the model and flexibility in the use of the model. For example, SELDM can be used to model runoff from various land covers or land uses by using the highway-site definition as long as representative water quality and impervious-fraction data are available.

Introduction

Water-resource managers are concerned about the frequencies, magnitudes, and durations of concentrations and loads (the products of measured stormflow and concentration) that may have an adverse effect on the quality of receiving waters (Driscoll and others, 1979, 1989; Athayde and others, 1983; Di Toro, 1984; U.S. Environmental Protection Agency, 1996b, 2002b, 2007a; Smith and others, 2001; Borsuk and others, 2002; Bonta and Cleland, 2003; Gibbons, 2003; Novotny, 2004; Elshorbagy and others, 2007; Brouwer and De Blois, 2008; Langseth and Brown, 2011). These decisionmakers commonly use specified estimates of streamflow and upstream constituent concentrations to estimate allowable concentrations and flows for discharges to receiving waters (U.S. Environmental Protection Agency, 1986b, 2002b). Evaluating the potential effects of stormwater, however, poses many unique challenges (Athayde and others, 1983; Di Toro, 1984). Intermittent and highly variable concentrations, flows, and loads complicate the monitoring, characterization, and evaluation of potential effects of runoff on receiving waters. For example, the U.S. Environmental Protection Agency's (USEPA) Nationwide Urban Runoff Program (NURP) evaluated the effects of short-term exposures that would result from intermittent stormwater runoff and estimated that acute concentrations in runoff would be about twice those for continuous discharges during steady low-flow conditions (Athayde and others, 1983). The NURP used event mean concentrations (EMCs) of constituents in runoff and receiving waters to evaluate the potential effects of runoff. EMCs are operationally defined as the total constituent load from a storm event divided by the total volume of runoff from the storm. EMCs are commonly estimated from data collected with flow-proportional water-quality-sampling methods. Planning-level estimates of EMCs in runoff and receiving waters at monitored sites can be used to evaluate the potential for adverse effects from highway and urban runoff in receiving waters at unmonitored sites (Athayde and others, 1983; Di Toro, 1984; Driscoll, Shelley and others, 1989; Driscoll and others, 1990b; Marsalek, 1991).

The Federal Highway Administration (FHWA) developed a highway-runoff model that used analytical approximations to estimate the potential effects of runoff on receiving waters. Publication of the 1990 FHWA runoff-quality model with data from 24 highway-runoff monitoring sites was the culmination of the FHWA runoff-quality research conducted during the 1970s and 1980s (Driscoll and others, 1990a, b). The 1990 FHWA runoff-quality model was based on this older runoff-quality data and the assumption that concentrations of constituents in receiving waters were equal to 0. By the mid-1990s, however, it was recognized that the existing data and modeling methods were reaching obsolescence (Bank and others, 1996). Changes in highway construction and maintenance (such as the use of pulverized rubber tires in pavement mixtures) and automobile technology (such as the disappearance of leaded fuel, continuing improvements in catalytic converters, and a trend from asbestos to organometallic brake pads) may affect the quality of highway runoff. Changes in atmospheric deposition and other ambient sources of pollution from surrounding land uses also could affect the quality of highway runoff and receiving waters. As a result of the implementation of Total Maximum Daily Load (TMDL) regulations, decisionmakers have become increasingly aware of the importance of considering the quality of upstream receiving waters in estimating the potential effects of runoff from highways and other land uses. Furthermore, awareness has been increasing that statistical approaches and Monte Carlo methods are needed to address the complexities that affect the probabilities of adverse effects from runoff because scientists, engineers, and decisionmakers now recognize the stochastic nature of stormflow variables, which are partly predictable and partly random. Thus, a model that could comprehensively incorporate new data and methods was needed.

The SELDM model was developed by the U.S. Geological Survey (USGS) in cooperation with the FHWA to supersede use of the 1990 FHWA runoff-quality model to indicate the risk for stormwater concentrations, flows, and loads to be above user-selected water-quality goals. The SELDM model was developed and tested during the 2009–12 period. SELDM is designed to be a tool that can be used to transform disparate and complex scientific data into meaningful information about the risk for adverse effects of runoff on receiving waters, the potential need for mitigation measures, and the potential effectiveness of such measures for reducing these risks. SELDM was designed to help inform water-management decisions for streams and lakes receiving highway runoff. Currently (2012), SELDM includes precipitation, streamflow, and water-quality data that are geographically referenced for sites in the conterminous United States. However, SELDM can be used for analysis of runoff quality in other areas by setting up geographically referenced datasets or by entering user-defined statistics for a site of interest.

Purpose and Scope

This report is a user's manual for the SELDM model. It provides information about the theory and implementation of the model including the Monte Carlo methods, the methods for defining hydrologic variables, numerical methods, and governing equations. It provides information for deriving model inputs and interpreting model outputs. It also provides a detailed discussion of the graphical user interface and the format of output files. Four appendixes provide modeling information. Appendix 1 describes numerical methods for Monte Carlo modeling. Appendix 2 describes specification of basin properties needed to characterize the highway site and the upstream basin. Appendix 3 provides an illustration of the database design. Appendix 4 provides step-by-step use of the program's graphical user interface.

SELDM was developed as a Microsoft Access® database software application that uses a simple graphical user interface to facilitate the storage, handling, and use of hydrologic datasets. The program is implemented within the database by using the Visual Basic for Applications® (VBA) programming language. Program source code for the analytical techniques is provided in SELDM and in electronic text files accompanying this report. Program source code that is specific to Microsoft Access®, the graphical user interface, and dataset handling is provided in the database. An installation package with a run-time version of the software is available with this report for potential users who do not have a compatible copy of Microsoft Access®. Administrative rights are needed to install this version of SELDM. The user needs full control (permission to read, write, and modify) of an output directory named "FHWA-SELDM" on the root drive of the computer to run the model. This directory must be distinct from the program-file directories used to install SELDM and the other programs developed to facilitate analysis of hydrologic and water-quality data (Granato, 2006, 2009, 2010; Granato and Cazenas, 2009; Granato and others, 2009).

Data-Quality Objectives

Data-quality objectives (DQOs) are criteria that are meant to ensure that data and interpretations are useful for the intended purpose (U.S. Environmental Protection Agency, 1986a, 1994, 1996a; Granato and others, 2003). DQOs are used to define the information and data necessary to develop credible estimates and make defensible decisions for managing environmental resources. The DQO process is designed to help evaluate the costs of data acquisition in relation to the consequences of a decision error caused by inadequate input data (U.S. Environmental Protection Agency, 1994, 1996a; Granato and others, 2003). SELDM is designed to facilitate an iterative DQO approach that is consistent with environmental risk-management methods used by the FHWA and the USEPA (Sevin, 1987; Cazenas and others, 1996; Federal Highway Administration, 1998; U.S. Environmental Protection Agency, 1996a).

The FHWA has established a system of water-quality-assessment and action plans represented by a decision tree that includes different levels of interpretive analysis to determine the potential environmental effects of highway runoff (Sevin, 1987; Cazenas and others, 1996; Federal Highway Administration, 1998). In the FHWA process, the state department of transportation (DOT) conducts an initial assessment to estimate the probability that the highway configuration being considered will produce unacceptable environmental effects. If the probable risk of an adverse effect is unacceptable to decisionmakers, the assessment is refined with more detailed data and analysis. The process is concluded if a low probability of unacceptable environmental effects can be demonstrated. The decision rule for DQOs in this process is dependent on the sensitivity of the receiving waters, the presence of water supplies in the watershed, uncertainties in available data, and limitations of the analysis (Patricia Cazenas, U.S. Department of Transportation, Federal Highway Administration, oral commun., 2005). The DOTs, however, commonly plan mitigation strategies to minimize the potential for adverse effects from highway runoff, even if criteria excursions are improbable (Patricia Cazenas, Federal Highway Administration, written commun., 2006). Water-quality excursions are defined herein as concentrations, flows, and loads in effluent or receiving waters that may cause or contribute to unacceptable environmental effects in receiving waters (U.S. Environmental Protection Agency, 1986a, 1994).

SELDM is designed to rapidly generate planning level estimates with available information and data and to refine such estimates if necessary. Planning-level estimates are defined as the results of analyses used to evaluate alternative management measures; planning-level estimates are recognized to include substantial uncertainties (commonly orders of magnitude) in all aspects of the decision process (Barnwell and Krenkel, 1982; Marsalek and Ng, 1989; Marsalek, 1991). To support a step-by-step refinement process, SELDM is designed to facilitate initial estimates based on available regional statistics determined by the location of the site of interest, to help refine statistics by selecting data from nearby hydrologically similar sites, and to accept user-defined statistics. User-defined statistics may be calculated from available data or data from monitoring studies done at the site

of interest as conditions warrant. Considerable uncertainty may remain, however, even if site-specific data are collected (Winter, 1981; Granato and others, 2003; Harmel and others, 2006; Smith and Granato, 2010; Granato, 2010, appendix 1).

Theory and Implementation

SELDM uses Monte Carlo methods to generate a stochastic population of the concentrations, flows, and loads needed to implement a mass-balance model for a receiving stream and (or) lake. SELDM also has a stochastic best management practice (BMP) module to assess the potential benefits of implementing stormwater controls at a site of interest. Monte Carlo methods are used because it is the combination of different distributions of precipitation, prestorm flows, runoff coefficients, and water-quality concentrations that determines the potential risk of water-quality excursions. Excursions are commonly associated with constituent concentrations that exceed a maximum allowable value, but for properties like pH, concentrations of dissolved oxygen, and streamflow, an excursion occurs if the values are below desirable limits. Deterministic methods are not able to characterize the interaction of different distributions for hydrologic parameters and BMP-performance measures. Unlike deterministic hydrologic models, SELDM is not calibrated by changing values of input variables to match a historical record of values. Instead, input variables for SELDM are based on site characteristics and representative statistics for each hydrologic variable. Each of these variables may be characterized by different probability distributions. Monte Carlo methods are needed because theoretical solutions depend too heavily on assumptions about the resultant distributions of concentrations, flows, and loads. The output results from SELDM, however, are not based on such assumptions. The benefit of the Monte Carlo analysis is not to decrease uncertainty in the input statistics, but to represent the different combinations of the values of variables that determine potential risks for water-quality excursions. Simpler methods may provide estimates of mean values, but it is commonly the extreme events that are of most interest to scientists, engineers, and decisionmakers for evaluating the potential for excursions.

A mass-balance approach (fig. 1) is commonly applied to estimate the concentrations and loads of water-quality constituents in receiving waters downstream of an urban or highway-runoff outfall (Warn and Brew, 1980; Di Toro, 1984; Driscoll and others, 1989; Driscoll and others, 1990b; Schwartz and Naiman, 1999). In a mass-balance model, the loads from the upstream basin and runoff source area (in this case, the highway) are added to calculate the discharge, concentration, and load in the receiving water downstream of a runoff discharge point. These models commonly are based on the assumptions that the runoff and the receiving water

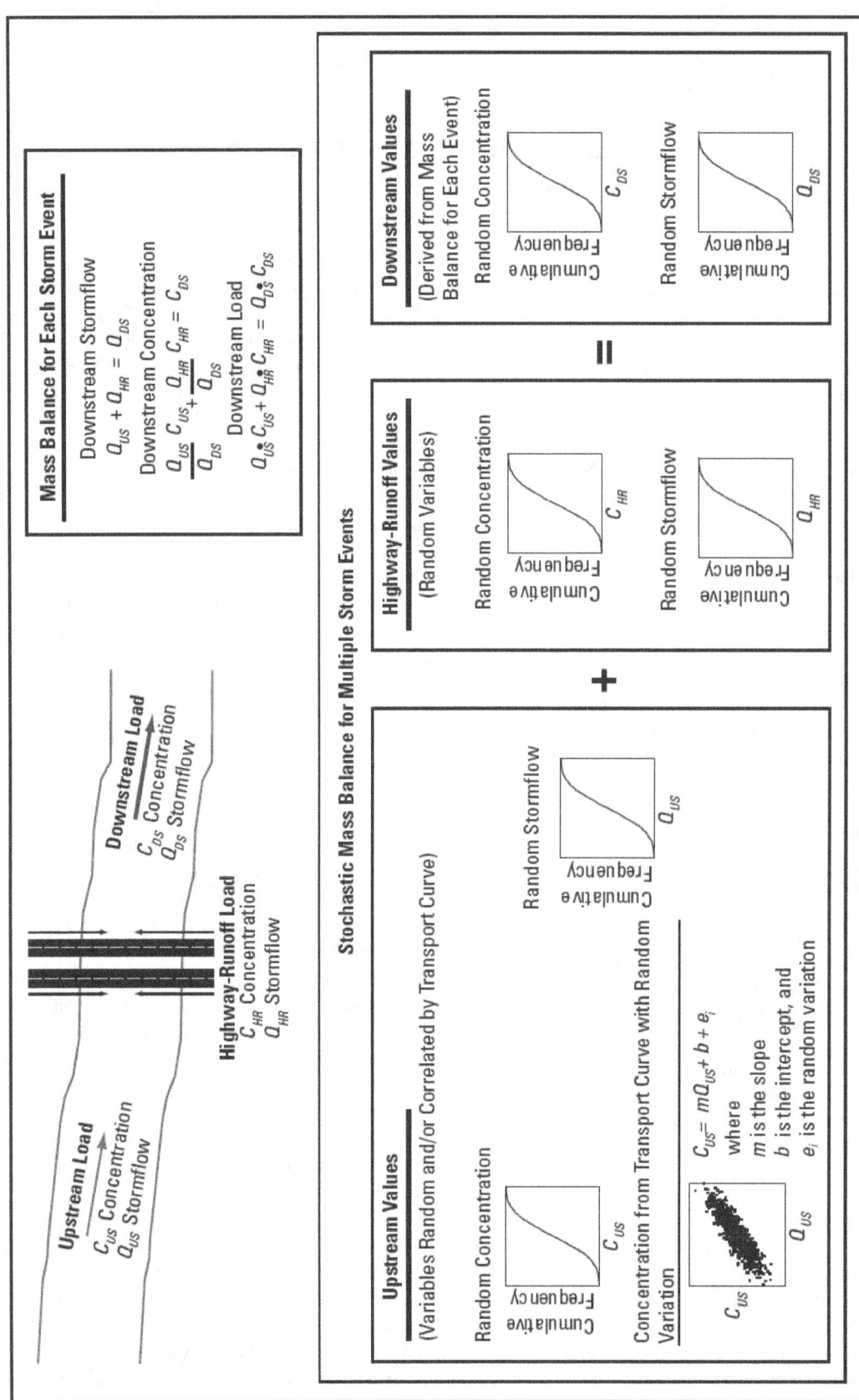

Figure 1.   The stochastic mass-balance approach for estimating stormflow, concentration, and loads of water-quality constituents upstream of a highway-runoff outfall, from the highway, and downstream of the outfall.

are fully mixed and that there are no reactions that reduce the mass of the constituent at the point of mixing.

The diagram in the upper left corner of figure 1 shows a hypothetical highway site crossing a stream with inputs from the upslope areas on both sides of the stream. The mass-balance calculation also may be done for the highway on each side of the stream and the bridge as separate contributing areas (one area at a time). A highway also may be parallel to the stream, in which case the mass-balance calculations could be modeled for each discharge point or for a conceptual discharge point incorporating the entire contributing area. The mass-balance approach also may be used to model any land use discharging runoff to a stream. In each case, the mass-balance equations in the upper right corner of figure 1 would be used to calculate the downstream stormflow, concentrations, and loads on the basis of runoff values from the selected areas.

Stochastic estimates of concentrations, stormflows, and loads of constituents are needed to use the mass-balance approach for estimating the potential for excursions in runoff and receiving waters (Warn and Brew, 1980; Schwartz and Naiman, 1999). Storm events commonly are defined as independent statistical events characterized by a volume, intensity, duration, and time between midpoints of successive storms for the purposes of planning, analysis, and sampling efforts (Driscoll and others, 1979; Athayde and others, 1983; Goforth and others, 1983; Driscoll, Palhegyi, and others, 1989; Driscoll, Shelley, and others, 1989; U.S. Environmental Protection Agency, 1992; Wanielista and Yousef, 1993; Adams and Papa, 2000; Church and others, 2003). Statistics describing the frequency distributions of component discharges and concentrations are needed to estimate the statistics for downstream discharges, concentrations, and loads. For example, Di Toro (1984) used information about probability distributions of stormflows and EMCs from the site of interest and the upstream basin to construct an empirical probabilistic dilution model to develop planning-level estimates of downstream EMCs and stormflow volumes. The lower half of figure 1 indicates how the random distributions of concentration and flow from the upstream basin and the highway site are used to calculate downstream values in SELDM.

Upstream concentrations may vary randomly or may be correlated with stormflows. For example, Di Toro (1984) based his method on the assumption that contributing stormflows and concentrations are independent and lognormally distributed. Warn and Brew (1980), however, indicated that upstream concentrations and loads are correlated. Schwartz and Naiman (1999) also demonstrated the effect of correlation between concentrations and flows in receiving waters on the adequacy of planning-level estimates of concentrations and loads in runoff. Granato and others (2009) developed water-quality transport curves, which are regression equations for estimating concentrations from streamflow. Their results indicate that concentrations of suspended sediment and sediment-associated constituents commonly increase with increasing streamflow, whereas concentrations of dissolved constituents such as total hardness commonly decrease with increasing streamflow. The resulting probability distributions of downstream EMCs indicate the potential for water-quality excursions and therefore the potential need for more information and data that may be used to identify suitable mitigation measures.

To calculate the concentrations, flows, and loads required for the mass-balance analyses (fig. 1), SELDM calculates values for 17 primary environmental variables, 15 of which are modeled as stochastic variables (table 1). For each storm, the volume of highway-runoff stormflow is calculated by using precipitation and highway-runoff-coefficient statistics. The timing of runoff from the highway is calculated as a function of site characteristics, a fixed hydrograph-recession ratio equal to 1, and storm duration. If BMP modifications to the highway runoff are specified, then the timing and volume of runoff from the BMP also is calculated. The volume of upstream stormflows is calculated by using prestorm-flow, precipitation, and runoff-coefficient statistics for the upstream basin. The timing of runoff from the upstream basin is calculated as a function of site characteristics, a stochastic hydrograph-recession ratio, and the storm duration. Dilution in the receiving water is calculated by using the volume of upstream flow that coincides with untreated highway runoff and the BMP discharge. The concentrations of upstream constituents are stochastic variables that can be calculated as purely random variables, dependent variables (defined as a function of another constituent), or as functions of upstream flow (the transport curve).

The annual lake-basin analysis also is a stochastic mass-balance model. The variables from the storm-by-storm highway-runoff calculations are added by annual-load accounting year to generate a population of annual loads. Loads from the rest of the lake basin, however, are the sums of loads associated with all daily streamflows in each annual-load accounting year, rather than just the stormflows, because dry-weather base flow can contribute substantially to annual loads. Although the concentration-attenuation factor is not a stochastic variable, the population of annual average lake concentrations is a stochastic variable because it is calculated as the sum of all influent loads divided by the sum of all influent flows for each annual-load accounting year.

Monte Carlo Simulation Methods

Monte Carlo simulation methods are modeling techniques that use random-number generators to repetitively draw random samples based on input statistics and then combine the different variables to determine the probability distribution for model outputs. Use of the term Monte Carlo for such simulations was coined during the Manhattan Project of World War II in a reference to the games of chance such as dice or the Roulette wheel that produce stochastic results (Saucier, 2000; Harrison, 2010). In hydrology, a stochastic process is commonly defined as a process with deterministic

Table 1.   Probability distributions used to model primary environmental variables in the Stochastic Empirical Loading and Dilution Model.

[Information on modeling of these distributions is in appendix 1. HRC, highway runoff concentration; HRQ, highway runoff flow; 2PExp, two-parameter exponential distribution; Norm, normal distribution; LogN, lognormal distribution (natural or common logarithms); PIII, Pearson type III distribution; LogPIII, log-Pearson type III distribution of the natural or common logarithms of the variable; Tri, triangular distribution; Trap, trapezoidal distribution]

Variable number	Environmental variable	Probability distributions								Censoring	Correlation
		Not stochastic	2PExp	Norm	LogN	PIII	LogPIII	Tri	Trap		
1	Precipitation volume		X							Yes	No
2	Precipitation duration		X							Yes	No
3	Time between storm-event midpoints		X							Yes	No
4	Prestorm flow		X[1]		X[2]		X[2]			No[3]	No
5	Upstream runoff coefficient			X[4]		X[4]				Yes	Variable 4
6	Upstream hydrograph recession							X[5]		Yes[6]	No
7	Highway runoff coefficient			X[4]		X[4]				Yes	Variable 5
8	Highway hydrograph recession	X								No	No
9	Water quality, random		X[1]	X[4]	X[2]	X[4]	X[2]			Yes[7]	No
10	Water quality, dependent			X[8]	X[8]					Yes[9]	Yes[10]
11	Water quality, transport curve			X[8]	X[8]					Yes[9]	Yes[10]
12	Water quality, downstream adverse effect ratio							X[5]	X[5]	Yes[6]	No
13	BMP, flow modification							X[5]	X[5]	Yes[6]	HRQ
14	BMP, hydrograph extension							X[5]	X[5]	Yes[6]	HRQ
15	BMP, concentration modification							X[5]	X[5]	Yes[6,11]	HRC
16	Lake, daily mean lake-basin streamflow				X		X			No	No
17	Lake, apparent annual concentration-attenuation factor	X								No	No

[1] A Pearson type III distribution with a skew of 2.0 is equivalent to an exponential distribution.

[2] A log Pearson type III distribution with a skew of zero is equivalent to lognormal distribution.

[3] The prestorm flows are not censored, but zero flows are modeled by using conditional probability methods.

[4] A Pearson type III distribution with a skew of zero is equivalent to normal distribution.

[5] The triangular distribution is equivalent to a trapezoidal distribution with equal upper and lower values of the most probable value.

[6] Censoring is accomplished by limiting the lower bound to values that are greater than or equal to zero.

[7] If the distribution is normal or Pearson Type III and the random concentration is less than or equal to zero, the value is fit within a lognormal lower tail with the specified mean and standard deviation.

[8] The stochastic portion of the variable is modeled by using normal (or lognormal) scatter around the regression equation.

[9] If the regression equation is untransformed and results in a concentration that is less than or equal to 0, then 0.002 is used.

[10] Rank correlation is not used, but correlation is implicit in the regression relations among variables.

[11] Censoring is accomplished by setting the minimum irreducible concentration to a value that is greater than or equal to zero.

and random components. For example, when runoff discharges to a stream, the sum of the two flow volumes is a deterministic calculation, but the flow volume from each source area results from the random combination of storm properties and the effects of antecedent conditions on this runoff from both areas. Similarly, a water-quality transport curve may indicate a deterministic relation between flow and concentrations (either dilution or washoff), but the data may show considerable scatter above and below the regression line. It may be said that many environmental variables would be deterministic if enough data were available, but such detailed descriptions would require a complete characterization on any appreciable scale.

Using computers to simulate random processes is difficult because computers, by design, are completely deterministic machines (Devroye, 1986; Press and others, 1992; L'Ecuyer, 1999; Saucier, 2000; Gentle, 2005; L'Ecuyer and Simard, 2007). Computers are useful tools because they will consistently produce the same results when given the same starting conditions. Therefore, computers produce pseudorandom numbers rather than actual random numbers. Pseudorandom numbers are produced deterministically, but they should be indistinguishable from a series generated by an actual random process (such as Brownian motion, radioactive decay, or the rolls of perfect dice or a perfect Roulette wheel). The measure of a random-number generator is the ability to produce such a series of numbers; many available generators do not pass such tests (Press and others, 1992; Hellekalek, 1998; L'Ecuyer, 1998; Marsaglia and Tsang, 2002; L'Ecuyer and Simard, 2007).

With SELDM, data can be modeled by using seven probability distributions (table 1), including the two-parameter exponential, the normal, the lognormal, the Pearson type III, the log-Pearson type III, the triangular, and the trapezoidal distribution (appendix 1). In some cases, one distribution is a special case of another distribution. For example, the exponential distribution is a Pearson type III distribution with a coefficient of skew equal to 2.0, and a normal distribution is a Pearson type III distribution with a coefficient of skew equal to 0. Similarly, the lognormal distribution is a log-Pearson type III distribution with a coefficient of skew (of the logarithms of data) of 0. The triangular distribution is a special case of the trapezoidal distribution with equal upper and lower bounds of the most probable value. The probability distributions commonly used to model each variable (Athayde and others, 1983; Di Toro, 1984; Driscoll, Palhegyi, and others, 1989; Driscoll and others, 1990b; Van Buren and others, 1997; Novotny, 2004; Vogel and others, 2005; Cheng and others, 2007; Granato and others, 2009; Granato, 2010) were selected for use in SELDM (table 1). In some cases—for example, the upstream hydrograph recession, the adverse-effects ratio, and the BMP-modification variables—no particular distribution is commonly used. In these cases, the triangular or trapezoidal distributions were selected for use in SELDM (table 1). These distributions were selected because they can be used to model these processes and are commonly recommended for selection

when expert judgment is used to model data (Haan, 1977; Johnson, 1997; Saucier, 2000; U.S. Environmental Protection Agency, 2001; Kacker and Lawrence, 2007). For example, the BMP-performance variables are ratios, and Johnson (1997) indicates that the triangular distribution is well suited to model ratios. Furthermore, BMP-performance statistics currently are highly uncertain, and these distributions are recommended for such cases (U.S. Environmental Protection Agency, 2001; Kacker and Lawrence, 2007).

Ten of the primary variables are generated independently, rank correlation coefficients can be specified by the user for four variables, the rank correlation coefficient is calculated by SELDM for one variable, and two variables may be correlated by using regression relations (table 1). Rank correlations can be specified between the upstream runoff coefficient and the prestorm flow volume, between the highway-runoff-flow volume and the BMP flow-modification and hydrograph-extension variables separately, and between the highway-runoff concentration and the BMP concentration-modification variable. The correlation between the highway-runoff coefficient and the upstream runoff coefficient is calculated by SELDM as a fixed function of the impervious fraction of each source area. The random concentration variables are not correlated to storm properties or flows in SELDM; the literature on highway- and urban-runoff quality indicates that such correlations are weak or nonexistent and do not have substantial effects on receiving-water concentrations (Warn and Brew, 1980; Athayde and others, 1983; Di Toro, 1984; Driscoll, Shelley and others, 1989; Driscoll and others, 1990b). However, the dependent and transport-curve concentration variables are correlated by specifying the regression relation and the variability of residuals.

SELDM uses the MRG32k3a combined multiple recursive random-number generator (CMRRNG) algorithm by L'Ecuyer (1999) to generate the uniform random numbers needed to do the Monte Carlo simulations (appendix 1). This algorithm was implemented in Visual Basic® (VB) for use with SELDM because the native random-number generators in Microsoft® VB and VBA used in the Microsoft Office® programs fail to meet basic standards for random number generators (L'Ecuyer and Simard, 2007; McCullough, 2008). The MRG32k3a generator produces a series of pseudorandom numbers by using the remainder of integer division (appendix 1). The initial values are known as the random seeds (Devroye, 1986; Press and others, 1992; L'Ecuyer, 1999; Saucier, 2000; Gentle, 2005; L'Ecuyer and Simard, 2007). MRG32k3a uses two initial seed values with preset coefficients to find the remainder of each seed and the associated modulus. The remainder values are then used to generate the next seed and so forth. The uniform random numbers between 0 and 1 are calculated by dividing the outputs by the modulus. A random-seed management algorithm was developed for SELDM to ensure that each runoff-quality analysis would be repeatable.

SELDM uses Monte Carlo methods (appendix 1) to model the variables and relations shown in table 1. The uniform random numbers are used as inputs to numerical algorithms for generating numbers that fit seven probability distributions. The program uses the inverse cumulative distribution function (CDF) method to generate numbers from exponential, triangular, and trapezoidal distributions (Haan, 1977; Saucier, 2000; Gentle, 2003; Kacker and Lawrence, 2007; Cheng and others, 2007). SELDM uses the frequency-factor method (Chow, 1954; Haan, 1977; Chow and others, 1988; Stedinger and others, 1993; Cheng and others, 2007) to generate numbers from normal, lognormal, Pearson Type III, and log-Pearson Type III distributions. SELDM uses the modified Wilson-Hilferty algorithm developed by Kirby (1972) to adjust the frequency factors to model Pearson Type III and log-Pearson Type III distributions. SELDM uses a modified frequency-factor method to generate random numbers to model regression relations. SELDM models rank correlation between selected variables by using an algorithm developed by Mykytka and Cheng (1994). SELDM models prestorm flows on intermittent or ephemeral streams that have a risk for zero prestorm streamflow by using random numbers that are adjusted for conditional probability.

Inverse Cumulative Distribution Function Method

The inverse CDF method (also known as the inverse transformation method) is a simple, efficient technique for generating random numbers from a specified probability distribution by using a set of uniform random numbers (Press and others, 1992; Saucier, 2000; Gentle, 2003; Cheng and others, 2007). If a random variable X has a CDF $F_X(X)$, then substituting the values of X will yield uniform random numbers in the range between 0 and 1 ($F_X(X)=U_{01}$). Thus, the inverse CDF can be used to generate values of X from U_{01} variate values ($F_X^{-1}(U_{01})=X$). This method is shown schematically in figure 2. Although the U_{01} values that are generated are evenly spaced within the interval from 0 to 1, the inverse CDF function controls the density of the output data. For example, the inverse CDF in figure 2 has low slopes in the tails and high slopes in the center of the distribution: the X values generated are tightly spaced in the center and sparse near the edges of the probability distribution function (PDF), even though the U_{01} values are evenly spaced. Implementations of the inverse CDF method commonly use the sample statistics within the algorithm to produce output values that meet the specified criteria, but a standardized distribution also may be used (Press and others, 1992; Saucier, 2000; Gentle, 2003).

SELDM uses the inverse CDF method with the two-parameter exponential distribution to generate stochastic data for the precipitation volume, duration, and time between storm-event midpoints (table 1). The two-parameter exponential distribution is modeled by using the user-selected minimum value and average value for each precipitation statistic. These statistics define the location and variability of the exponential values that are generated as described in appendix 1. The minimum value has the effect of censoring

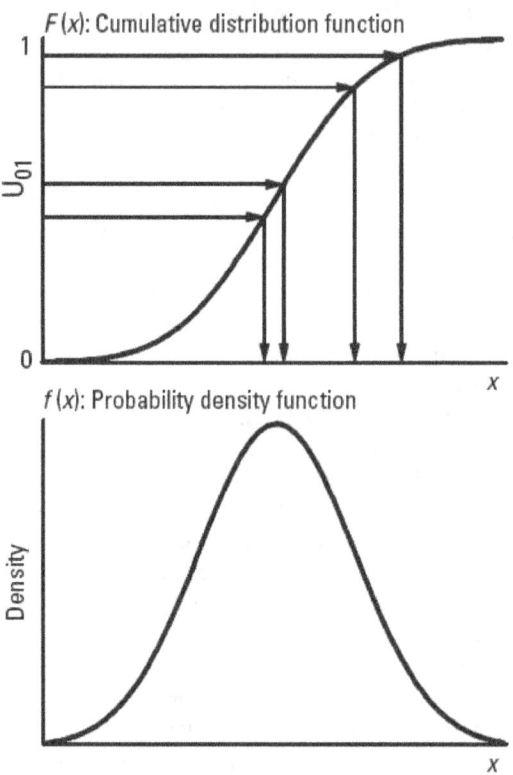

Figure 2. The inverse cumulative distribution function method for generating data that fit a given distribution (Modified from Saucier, 2000).

values from a two-parameter exponential distribution because the inverse CDF does not include values below the minimum. The time between storm-event midpoints is generated independently from the event duration and is used only to delineate annual-load accounting years and therefore the total number of storms generated (appendix 1).

SELDM uses the inverse CDF method with the triangular/trapezoidal family of distributions to generate stochastic data for the upstream hydrograph-recession variable, the adverse-effect ratio, and the three BMP-treatment variables (table 1). The upstream hydrograph-recession variable is limited to the triangular distribution because it is defined by the minimum value, most probable value, and maximum value specified by the user. The other four variables can be specified by using a trapezoidal distribution. These variables are defined by the minimum value, the lower bound of the most probable value, the upper bound of the most probable value, and maximum value specified by the user. If the lower and upper bounds of the most probable value are specified as being equal, then SELDM will produce stochastic data that fit the triangular distribution. These variables are generated using an algorithm developed by Kacker and Lawrence (2007), which is described in appendix 1.

The Frequency-Factor Method

The frequency-factor method is used to model distributions with inverse CDFs that are mathematically intractable, which means that they cannot be solved analytically by using elementary mathematical functions (Chow, 1954; Haan, 1977; Chow and others, 1988; Stedinger and others, 1993; Cheng and others, 2007). In both the frequency-factor method and the inverse CDF method, a random set of U_{01} values is transformed to a set of data values by using an algebraic expression that generates the correct probability density (fig. 2). The frequency-factor method is an efficient technique for generating random numbers from a specified probability distribution with a set of U_{01} variates (appendix 1). Chow (1954) showed that the frequency-factor method can be used to calculate values of a variable using the average, standard deviation, and standardized variate of a specified distribution by using the equation:

$$X_d = \bar{X} + S \times K_d, \qquad (1)$$

where

$\quad X_d \quad$ is a value from distribution d,

$\quad \bar{X} \quad$ is the average value used to generate stochastic data,

$\quad S \quad$ is the standard deviation used to generate stochastic data, and

$\quad K_d \quad$ is the variate associated with the value X_d for the selected distribution (d).

The skew of the data can be modeled by selecting a standardized form of a skewed probability distribution. The relation between the probability of occurrence and K_d depends on the distribution being modeled. K_d values can be generated from the U_{01} variates by using an interpolation table or algebraic approximations for the inverse CDF; SELDM uses algebraic approximations for the inverse CDF. The AS–241 algorithm by Wichura (1988) is used to generate normal variates K_N from the U_{01} values, which are generated by using the MRG32k3a algorithm. SELDM uses the modified Wilson-Hilferty transformation algorithm developed by Kirby (1972) to generate Pearson Type III variates K_P from normal variates K_N, which are generated by using the MRG32k3 and AS–241 algorithms. The lognormal and log-Pearson Type III variables are generated by using the K_N and K_P variates, which are based on statistics for the logarithms of the data, and then retransforming each value.

The frequency-factor method is used directly to generate five primary hydrologic variables in SELDM (table 1). The prestorm flow and daily mean lake-basin streamflow variables can be modeled as lognormal variables if the skew of the logarithms of streamflows is specified as being equal to 0. These variables will be modeled as log-Pearson Type III variables if the skew of the logarithms of streamflows is specified as not being equal to 0. Similarly, the runoff coefficients can be modeled as normal or Pearson Type III variables depending on the specified skew value. The runoff coefficients, however, are censored by using standard acceptance-rejection methods (Press and others, 1992; Saucier, 2000; Gentle, 2003) so that values that are less than or equal to 0 or greater than 1 will be rejected. Random water-quality variables may be specified by using statistics for the normal, Pearson Type III, lognormal, or log-Pearson Type III distributions; values less than or equal to 0 are replaced.

Stochastic Regression Method

The stochastic regression method uses statistics defining a linear regression equation and the scatter of data above and below the regression line to generate data that represents the relation between measured data values for the variables of interest. Linear regression analysis is the process of fitting a straight line to a set of data (or some transformation of the data) to obtain a mathematical expression for estimating the mean value of the dependent variable from a given value of the independent variable (Haan, 1977; Helsel and Hirsch, 2002). Two of the primary assumptions that underlie regression analyses are that the residuals are normally distributed with a local mean value centered on the line, and that the residuals have constant variance within the range of the independent variable characterized by the line. Given this, the stochastic-regression method uses the frequency-factor method to calculate values of dependent variable above or below the regression line. The regression equation with this stochastic component is

$$Y_i = b + m \times X_i + (KN_i \times \sigma_r), \qquad (2)$$

where

$\quad Y_i \quad$ is the ith value of the dependent variable,

$\quad b \quad$ is the intercept of the regression line,

$\quad m \quad$ is the slope of the regression line,

$\quad X_i \quad$ is the ith value of the independent (or predictor) variable,

$\quad KN_i \quad$ is the ith value of the random normal variate, and

$\quad \sigma_r \quad$ is the standard deviation of the residuals from the regression analysis.

The intercept, slope, and standard deviation of the residuals are input by the user, presumably from analysis of available water-quality data. As described in appendix 1, the standard deviation of the residuals (σ_r) and a normal frequency factor calculated from a uniform random variate (U_{01}) are used to determine the placement of the generated data point (Y_i) above or below the local regression-line value calculated on the basis of X_i. The standard deviation of residuals (σ_r) is characterized by the median absolute deviation (MAD) of residuals, which is a nonparametric measure of σ_r (Helsel and Hirsch, 2002; Granato, 2006; Granato and others, 2009). The MAD was selected to represent the variability of residuals without the undue effect that outliers may have on the σ_r value. The independent variable is another stochastic water-quality variable, or for a transport curve, the stochastic sample of upstream stormflows, which are calculated as the sum of prestorm flow and runoff averaged over the storm event (table 1). The SELDM interface provides several options for defining stochastic regression relations, including regression relations with one, two, or three segments and regression relations based on logarithmic transformations of the independent and dependent variables. The Kendall-Thiel Robust line (KTRLine) program (Granato, 2006) developed for the SELDM project can be used to calculate these regression statistics for a one-segment or a multisegment model. Because the multisegment models generated by the KTRLine program are based on nonparametric statistics, the potential for gross overestimation or underestimation of constituent concentrations is minimized within the limits of available data (Granato, 2006; Granato and others, 2009).

Regression relations in SELDM may include one to three segments each with positive, negative, or zero slopes and segment-specific MAD values. If the population of residuals is normally distributed, the regression-line prediction of the median Y_i value will approximate the mean of Y_i values for a given X_i, and the MAD will approximate the standard deviation (σ_r) of the Y_i values at that point within the interval represented by a given segment. If there is a linear one-segment relation between the predictor (X_i) and response (Y_i) variables, the slope of the line will be significantly different from 0 (either positive, as in figure 3A, or negative), and the population of Y_i values will have a normal distribution of data above and below the line. Thus, these data vary with X and have a random error component. If the slope is not significantly different from 0 (fig. 3B), the X_i term in equation 2 will drop out, the intercept of the line will represent the median and mean of Y_i, and the population of values will have a normal distribution of data above and below the intercept. The data are random with respect to X and are described only by the Y-population statistics, which are characterized by the intercept and error components of the regression analysis. If multiple processes affect relations between the predictor (X_i) and response (Y_i) variables, different relations might predominate over different ranges of the predictor variable. Figure 3C shows the case in which there is random variation in the response variable (for example, base-flow conditions) until a second process predominates (for example, runoff). In this case, each segment will have a different slope, intercept, and MAD. Granato and others (2009) provide several examples of the development and use of regression relations with a stochastic component. The water-quality transport curves for suspended-sediment concentrations, total phosphorus, and total hardness developed by Granato and others (2009) have been preloaded into SELDM for each ecoregion.

Logarithmic regression relations may better reflect the characteristics of hydrologic data (Helsel and Hirsch, 2002; Granato, 2006; Vogel and others, 2005). The use of logarithmic regression relations also precludes generation

A. One-segment regression model with a positive slope

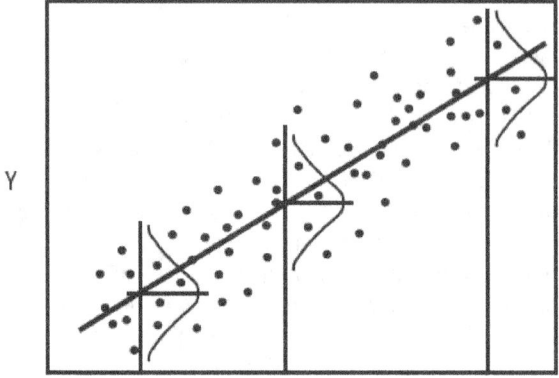

B. One-segment regression model with a zero slope

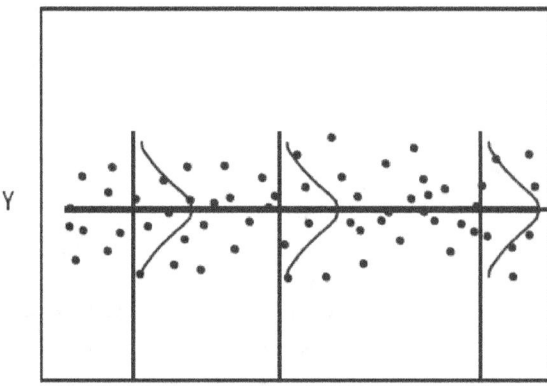

C. Two-segment regression model with a zero slope and a positive slope

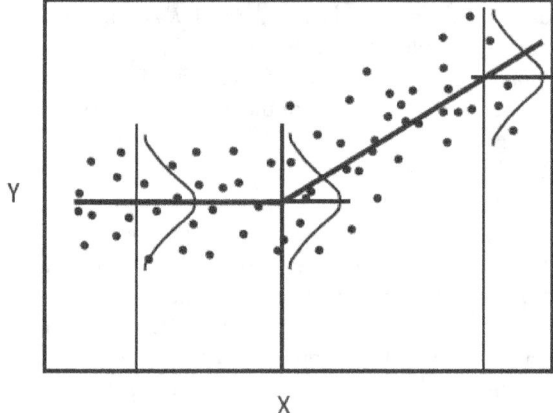

Figure 3. The normal distribution of data above and below the regression lines for *A,* a one-segment regression model with positive slope, *B,* a one-segment regression model with zero slope, and *C,* a two-segment regression model (Modified from Granato and others 2009, p. 16).

of data that are less than or equal to 0. SELDM does not reject values that are less than 0, so the user must take care to specify untransformed regression statistics that will not produce concentration values less than 0. If a concentration that is less than or equal to 0 is generated, then SELDM resets the concentration to equal the arbitrarily selected value of 0.002. This process may bias output results if the number of generated values less than or equal to 0 is substantial. Thus, output from SELDM that includes multiple concentration values equal to 0.002 may indicate that the input statistics produce concentration values less than or equal to 0.

Correlated Random Numbers

Correlation analysis is a method to quantify the type and strength of relations between two variables (Haan, 1977; Press and others, 1992; Helsel and Hirsch, 2002). SELDM uses the rank correlation coefficient known as Spearman's rho (ρ) to model statistical relations between variables. Spearman's rho is calculated by ranking the data and calculating the correlation coefficients between the rank values rather than the data values (Haan, 1977; Helsel and Hirsch, 2002). Spearman's rho indicates the strength of the relation regardless of the linearity of the relation between variables. Correlation coefficients commonly are dimensionless and are scaled to be in the range of -1 to 1, inclusive. The sign of the correlation coefficient indicates the type of relation. A positive sign indicates that one variable generally increases as the other increases; a negative sign indicates that one variable generally decreases as the other increases; and a value of 0 indicates that variations in the two variables are totally unrelated. The strength of relations between two variables is indicated by the value of the correlation coefficient. The relation between variables goes from random association (no relation) to monotonic covariance as the absolute value of the correlation coefficient increases from 0 to 1. For example, figure 4 shows the results of eight Monte Carlo experiments, each using 500 paired uniform random numbers. The results of the experiments demonstrate the amount of scatter that may be associated with different correlation coefficient values. The diagonal line in each graph indicates the perfect one-to-one relation that would be evident if the correlation coefficient were equal to 1. The graphs indicate the increasing scatter in

the paired values as the rank correlation coefficient, in this case Spearman's rho (ρ), decreases. It is notable that there is considerable scatter in the relation on almost all of the graphs, even on the graph representing a rho value as high as 0.85. Furthermore, the relation between variables appears to be almost random in the graph representing a rho value as high as 0.45 (absent the 1:1 line). Relations between variables for negative rho values would show the same scatter, but the trend would be from the upper left to the lower right in each graph. The statistical significance of the correlation coefficient is a function of the absolute value of the coefficient and the sample size. Figure 4 indicates that it is more difficult to distinguish between actual correlation and an accidental correlation caused by random sampling as the absolute value of the correlation decreases, even with large sample sizes.

In SELDM, four variables are calculated by using rank correlation to other variables (table 1). The upstream runoff coefficients are generated by using the rank correlation to the prestorm flow; the highway-runoff coefficients are generated by using the rank correlation to the upstream runoff coefficients; the BMP flow-modification and hydrograph-extension variables are generated by using the rank correlation to the highway-runoff inflow volumes; and the BMP-concentration-modification variables are generated by using the rank correlation to the highway-runoff inflow concentrations. The rank correlation between the highway- and upstream-runoff coefficients is calculated by SELDM, but the rank correlations for all the other variables are user defined.

Stochastic data generated from a specified rank correlation between the primary U_{01} variate (XU) and the secondary U_{01} variate (YU) are calculated by using an algorithm developed by Mykytka and Cheng (1994). Implementation of this algorithm is described in appendix 1. The secondary variate is calculated from the primary variate and an intermediate uniform random variate as a function of the rank correlation coefficient. The correlation of the uniform random variates is equal to the rank correlation of these variates because the ranks are the product of the value of the variate plus a plotting-position adjustment and the number of variates to be generated plus a plotting-position adjustment. The generalized equation for producing correlated uniform variates is

$$Y_3U_i = f\left[\rho' \times XU_i + \sqrt{1-(\rho')^2} \times Y_2U_i + 0.5 \times \left(1-\rho'-\sqrt{1-(\rho')^2}\right)\right], \tag{3}$$

where

Y_3U_i	is the ith value of the output U_{01} variate,
ρ'	is the absolute value of the correlation coefficient,
XU_i	is the ith value of the input U_{01} variate,
Y_2U_i	is the ith value an intermediate U_{01} variate, and
$f[\ ]$	is a transformation function that depends on the value of Y_2U_i and the sign of ρ.

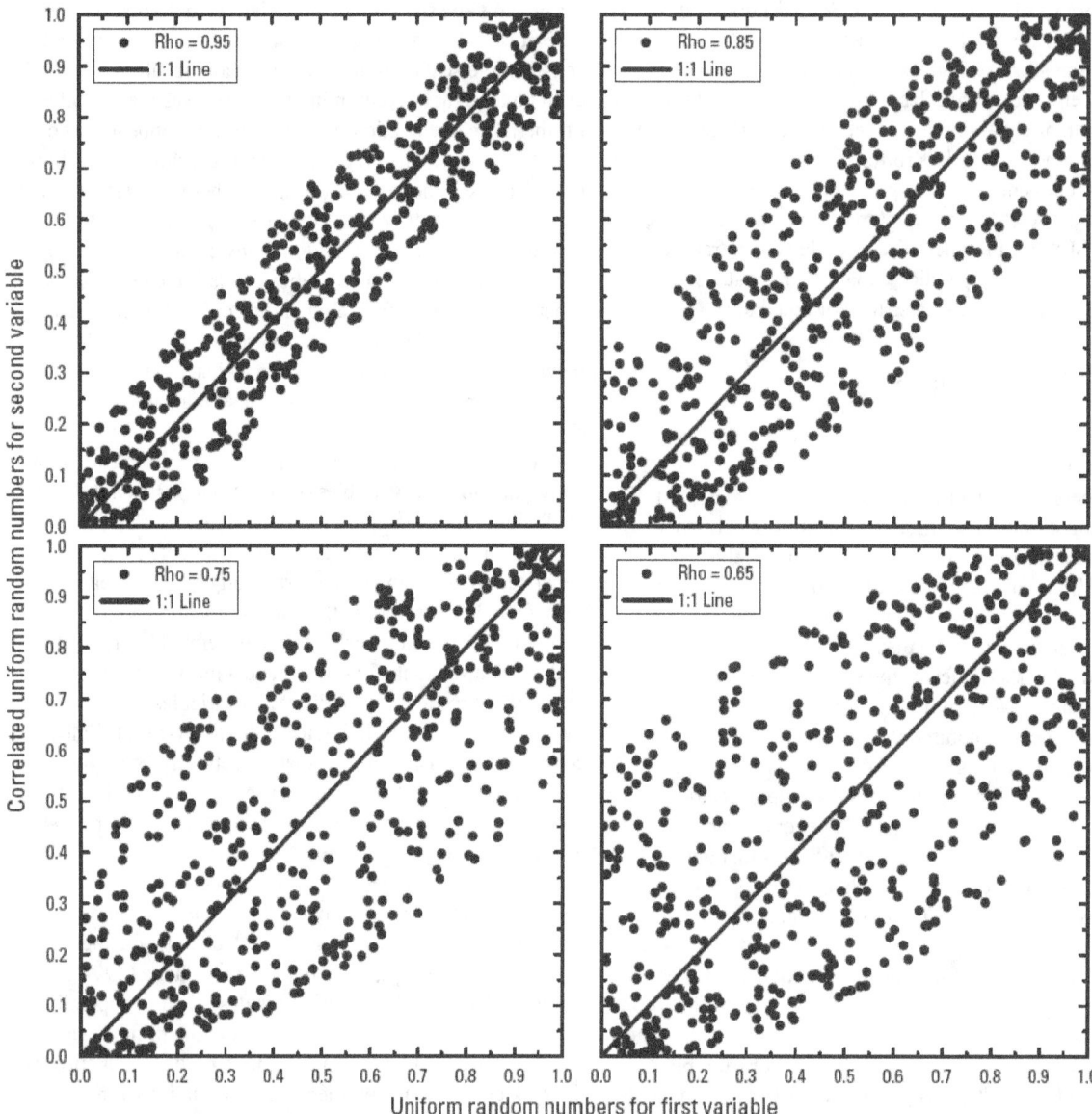

Figure 4. The results of a Monte Carlo analysis to demonstrate scatter of paired uniform random-number samples around a one-to-one relation for eight different values of the rank correlation coefficient (Spearman's rho). Each sample consists of 500 paired uniform random numbers in the range between 0 and 1.

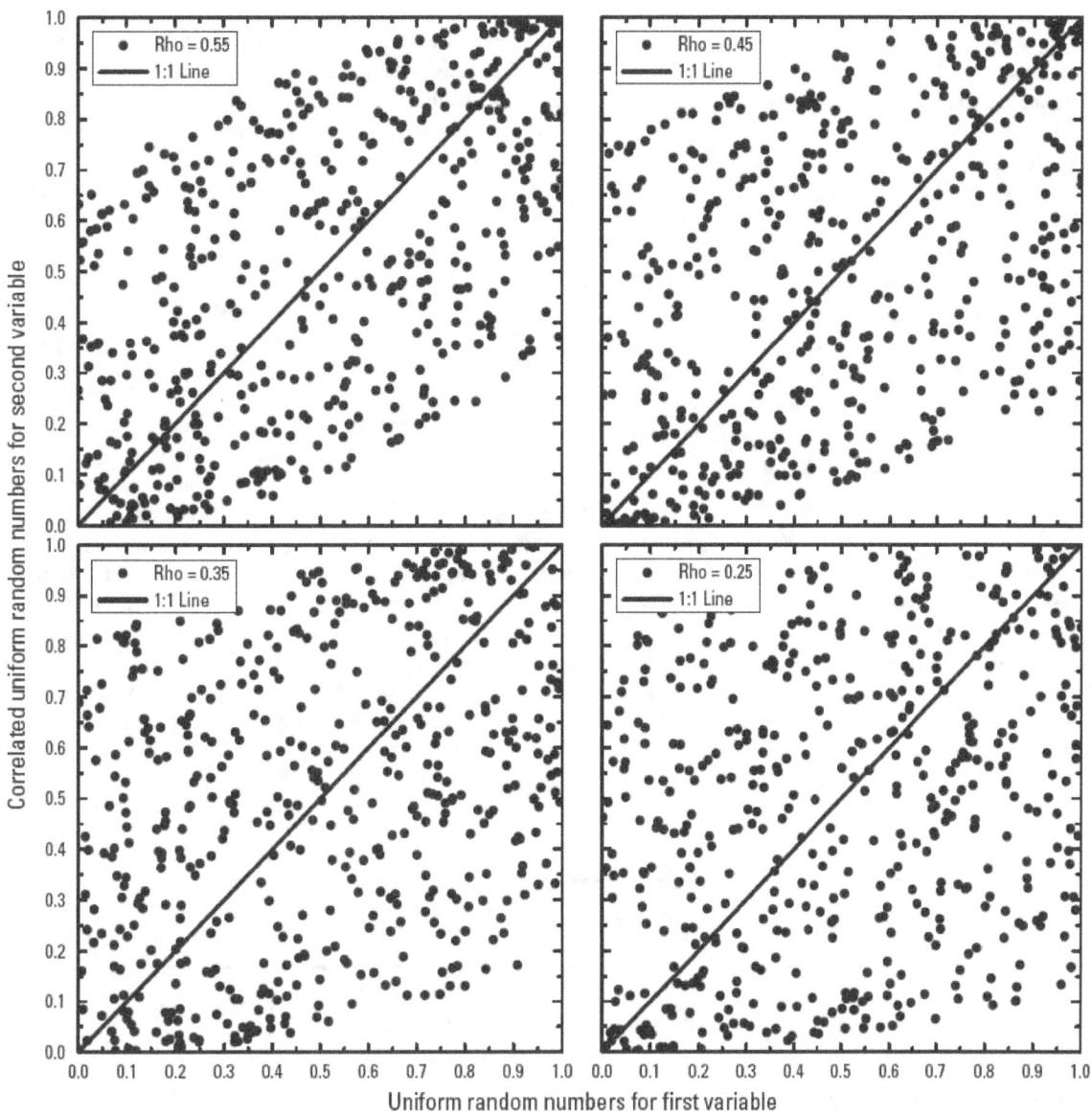

Figure 4. The results of a Monte Carlo analysis to demonstrate scatter of paired uniform random-number samples around a one-to-one relation for eight different values of the rank correlation coefficient (Spearman's rho). Each sample consists of 500 paired uniform random numbers in the range between 0 and 1.—Continued

The algorithm developed by Mykytka and Cheng (1994) was tested for use in SELDM by generating 141,282 sets of 500 and 1,000 U_{01} variates. The mean, standard deviation, output correlation coefficient, and the probability plot correlation coefficient (PPCC) statistics (Vogel and Kroll, 1989) were calculated for each set of output values. These tests showed that the algorithm developed by Mykytka and Cheng (1994) preserves the theoretical mean (0.5), the theoretical standard deviation (about 0.288675), and the theoretical marginal distribution of the U_{01} variates with small random variations that are within theoretical tolerance limits for the sample sizes generated. On average, the algorithm produces a small degree of bias dependent on the absolute value of ρ in the population of output ρ values, but this can be overcome by adjusting the input ρ value before doing the rest of the calculations (Mykytka and Cheng, 1994).

Stormflow

Estimates of stormflows are needed to use a mass-balance approach for predicting the stormflows, concentrations, and loads of constituents of concern in runoff and receiving waters (Warn and Brew, 1980; Di Toro, 1984; Schwartz and Naiman, 1999; Granato, 2010). Highway and urban runoff-quality assessments are based on storm-event analyses to characterize the potential effects of stormwater discharges on receiving waters. The mass-balance approach for storm-event analyses is based on estimates of upstream- and highway-runoff discharges. The total upstream stormflow component for each storm event comprises prestorm streamflow and the upstream storm runoff. This runoff is the product of the storm-event characteristics, the drainage area, and the volumetric runoff coefficient, which is defined as the ratio of runoff to precipitation volume (fig. 5). Similarly, the highway-runoff discharge for each storm event is the product of the storm-event characteristics, the drainage area, and the volumetric runoff coefficient. The storm-event characteristics and the runoff coefficients are stochastic variables in SELDM.

The relative importance of each stormflow component in determining downstream stormflow, concentrations, and loads depends on storm-event characteristics, upstream-basin characteristics, highway-catchment characteristics, and BMP characteristics. At one extreme, runoff from a highway catchment may contribute all of the downstream flow from a small storm during which rainfall is completely absorbed by soils in a pervious rural basin with an ephemeral stream. At another extreme, runoff from a highway catchment in a large basin with a large perennial stream may cause undetectable changes in downstream stormflow and water quality. The

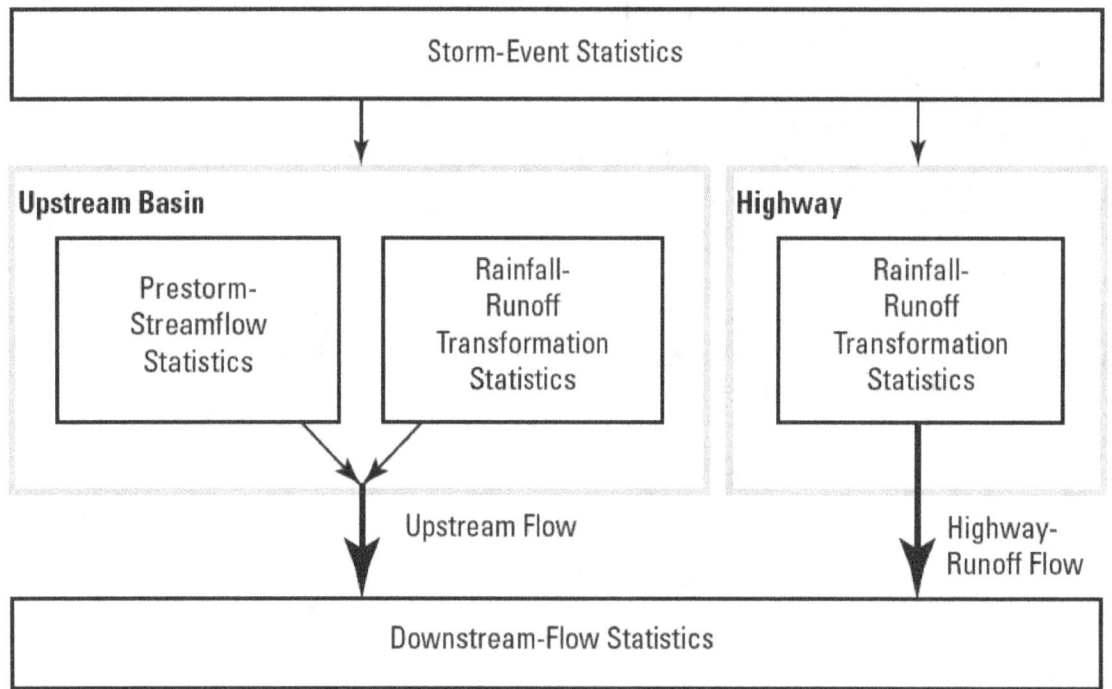

Figure 5.   The upstream-flow and highway-runoff components that must be estimated for a mass-balance analysis of receiving-water quality.

general geographic location is important because streamflow statistics, which determine prestorm flows, and storm-event statistics vary spatially across the United States. The hydrologic basin characteristics determine the volume and timing of the runoff component. For example, volumetric runoff coefficients are a function of the fraction of total impervious area (TIA).

Estimates of the timing of runoff from the highway site and the upstream basin are also needed to estimate potential effects of runoff in receiving waters. The timing of runoff is a function of the length, slope, and drainage features of the highway site and the upstream basin (methods for determining these variables are described in detail in appendix 2). These drainage features, which include the proportion of curb and gutter streets, storm sewers, and engineered drainage improvements, are characterized by the Basin Development Factor (BDF) developed by Sauer and others (1983). Specifically, estimates of upstream stormflows during the period when highway runoff is discharging to the stream are needed to calculate the potential dilution of highway runoff in the receiving stream. The duration of highway runoff determines the proportion of the upstream stormflow hydrograph that contributes to the concurrent downstream flow. Therefore, BMPs that extend the duration of the highway-runoff hydrograph within or beyond the upstream stormflow hydrograph may increase the total concurrent flow and thus the dilution of runoff constituents in the receiving water.

SELDM is a lumped parameter model because the highway site, the upstream basin, and the lake basin each are represented as single homogenous units. Results from SELDM are calculated as point estimates at the site of interest. Each of these source areas is represented by average basin properties. For example, highway-runoff results are produced for the highway site and the outlet of a BMP, if a BMP is specified. Upstream-basin results are produced for the point at which the highway runoff enters the stream. Use of the lumped parameter approach provides a number of advantages over distributed modeling approaches. First, use of the lumped parameter approach facilitates rapid specification of model parameters to develop planning-level models. Second, this approach also is representative of the detail available in most datasets. For example, available datasets for highway runoff and BMP performance commonly are limited to a small number of representative sites. Similarly, watershed studies for stream- or lake-quality monitoring commonly are based on data collected at sites that represent multiple land covers and different tributaries in a single drainage basin. Third, this approach allows for parsimony in the required inputs to and outputs from the model. Fourth, this approach allows flexibility in the use of SELDM. Because runoff from the highway site is calculated with only 5 basic hydraulic basin properties (appendix 2), SELDM can be used to model runoff from areas with other land uses, including the natural predevelopment land use, residential land uses, commercial land uses, or mixed land uses.

Storm-Event Characteristics

SELDM uses Monte Carlo methods to calculate the precipitation volume, precipitation duration, and the time to the next storm-event midpoint for each storm as a stochastic variable (appendix 1). Granato (2010) provides a detailed discussion of the methods and data for estimating storm-event characteristics for use with SELDM. The statistics available in SELDM for storm-event characteristics were based on data from the 2,610 selected National Weather Service hourly-precipitation data stations in the conterminous United States shown in figure 6 (Granato, 2010). The synoptic statistics characterize each storm as a discrete event over the entire basin without regard to within-storm variations. These storm events are characterized by the minimum interevent time (IET), total event duration, total event volume, and interval between storm-event midpoints (fig. 7). This definition of a storm event commonly is used for planning-level estimates of the quantity and quality of highway and urban runoff, the design and evaluation of runoff-quality BMPs, and the simulation of runoff flows (Driscoll and others, 1979; Goforth and others, 1983; Adams and others, 1986; Strecker, Mayo, and others, 2001; Driscoll, Palhegyi, and others, 1989; Driscoll and others, 1990a, b; U.S. Environmental Protection Agency, 1992; Adams and Papa, 2000; Granato, 2010).

SELDM uses the two-parameter exponential distribution to stochastically generate the duration, volume, and time between storm-event midpoints (table 1; appendix 1). SELDM generates uniform random numbers between 0 and 1 for each storm-event variable and uses these values with the inverse CDF of the two-parameter exponential distribution. Granato (2010) examined the use of several common probability distributions for modeling storm-event characteristics: the one-parameter exponential, the two-parameter exponential, the two-parameter lognormal, the two-parameter gamma, and the Pearson type III distribution. The two-parameter exponential distribution was selected because it preserves the characteristics of input statistics and is an efficient method for generating stochastic storm events. The two-parameter exponential distribution is parameterized by the mean and minimum values; the standard deviation and skew are functions of these values. The minimum volume of 0.1 inch (in.) and the minimum IET of 6 hours are the values defined by the USEPA as a runoff-generating event (Driscoll, Palhegyi, and others, 1989; U.S. Environmental Protection Agency, 1992; Granato, 2010). Use of the hourly data fixes the minimum storm duration at 1 hour, and the minimum time between storm-event midpoints (7 hours) is 1 hour plus the IET. This approach is consistent with the synoptic statistics calculated for most stations in the dataset. These values are the default in SELDM, but may be changed by loading a new dataset or by using the user-defined statistics option on the synoptic storm-event precipitation-statistics form.

Each storm that is generated for an analysis is identified by sequence number and annual-load accounting year. The model generates each storm randomly; there is no serial

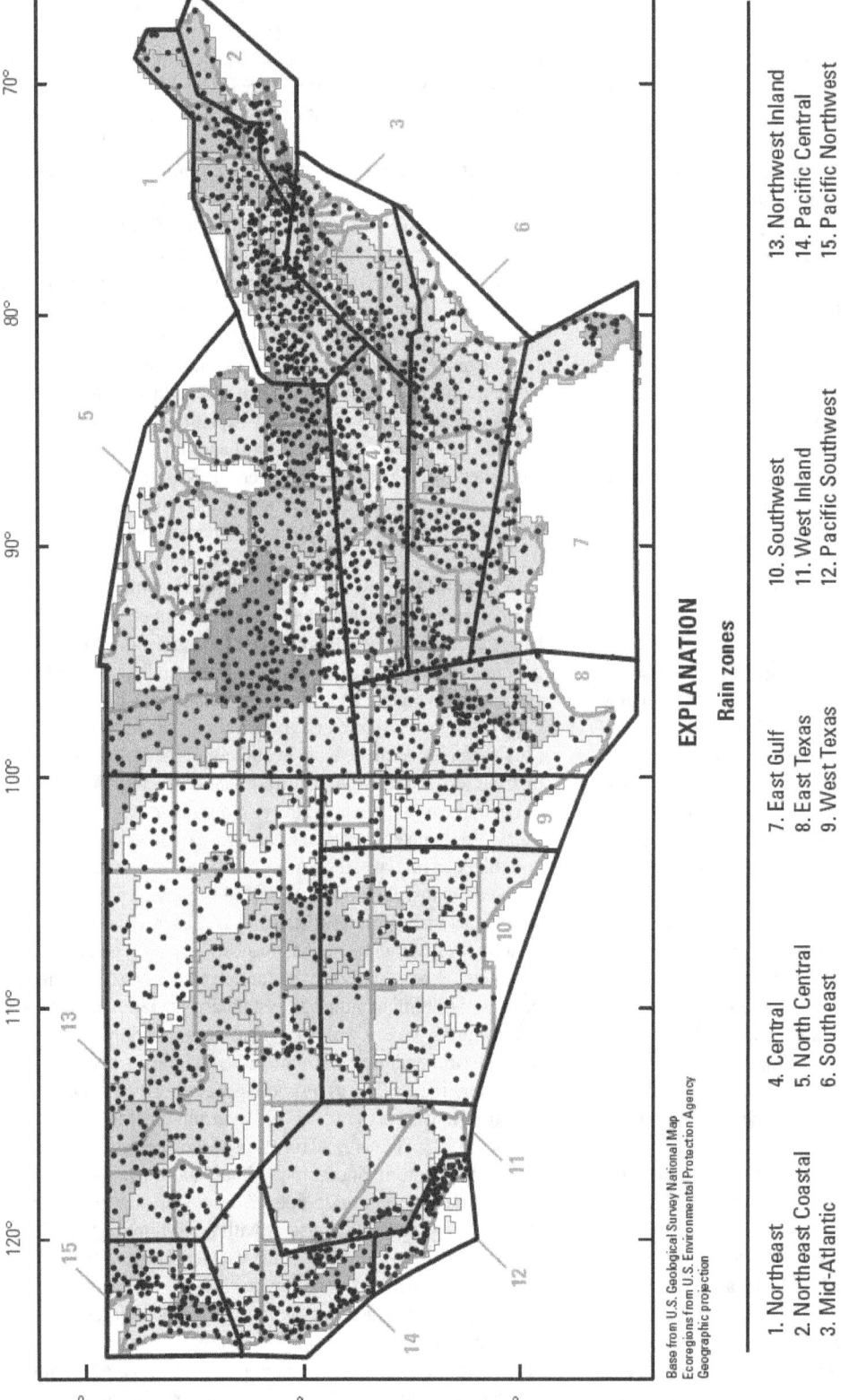

Base from U.S. Geological Survey National Map
Ecoregions from U.S. Environmental Protection Agency
Geographic projection

EXPLANATION

Rain zones

1. Northeast	4. Central	7. East Gulf	10. Southwest	13. Northwest Inland
2. Northeast Coastal	5. North Central	8. East Texas	11. West Inland	14. Pacific Central
3. Mid-Atlantic	6. Southeast	9. West Texas	12. Pacific Southwest	15. Pacific Northwest

Figure 6.   The spatial distribution of 2,610 hourly-precipitation data stations (black dots) with respect to the U.S. Environmental Protection Agency rain zones (Driscoll, Palhegyi, and others, 1989) and the U.S. Environmental Protection Agency (2003) Level III ecoregions (colored polygons) that have been discretized to a 15-minute grid in the conterminous United States. Ecoregions are identified on the plate useco.pdf on the CD–ROM accompanying this report.

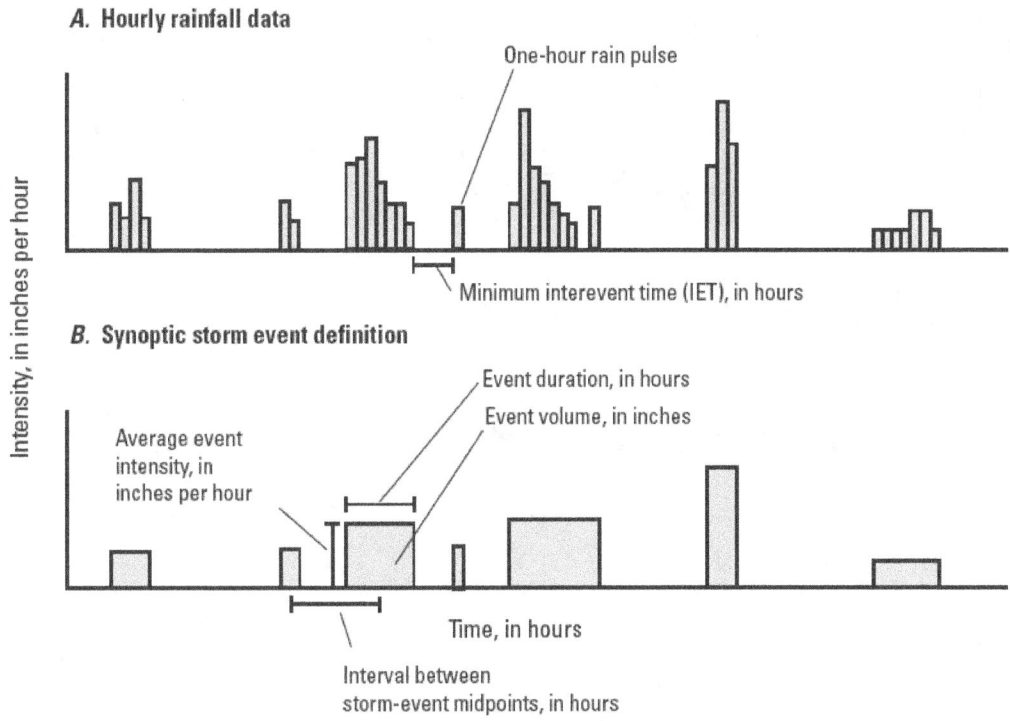

Figure 7. Schematic diagram showing A, The characterization of hourly rainfall data, and B, the synoptic storm-event definition (Modified from Driscoll, Palhegyi, and others, 1989).

correlation. The order of storms does not reflect seasonal patterns. The annual-load accounting years, which are random collections of events generated with sums of interevent times less than or equal to a year, are used to generate annual highway flows and loads for both the TMDL and the lake-basin analyses. Granato (2010) checked the results of the random-number algorithm and verified that use of the two-parameter exponential distribution would produce the correct number and distribution of storms and annual precipitation volumes for the selected minimum precipitation volume and IET. Selection of runoff-generating events with a minimum volume of 0.1 in. and a minimum IET of 6 hours is expected to reduce the number of storms and annual precipitation totals because a considerable number of very small storms are not included, and some storms become aggregated into larger and longer storms (fig. 7) (Driscoll, Palhegyi, and others, 1989; Adams and Papa, 2000; Granato, 2010).

SELDM offers seven options for selecting storm-event statistics on the synoptic storm-event-precipitation statistics form. These options include selecting (1) the mean of station statistics by USEPA rain zone, (2) the median of station statistics by USEPA rain zone, (3) the mean of station statistics by ecoregion, (4) the median of station statistics by ecoregion, (5) the mean of statistics from user-selected stations, (6) the median of statistics from user-selected stations, or (7) entering user-defined statistics. Granato (2010) selected the USEPA rain zones and USEPA Level III nutrient ecoregions to provide regional planning-level estimates of precipitation statistics because these regions are accepted for water-quality monitoring and management. Methods for selecting these options on the statistics form are described in detail in appendix 4. The default rain zone and ecoregion are automatically selected by entering the latitude and longitude of the highway site. The user, however, can manually select an ecoregion that better represents conditions at a site of interest. Precipitation statistics can be calculated for the site of interest from the mean or median of statistics from selected hourly-precipitation data stations at nearby hydrologically similar sites. The option for entering user-defined statistics can be used to enter site-specific statistics, do a sensitivity analysis, or evaluate the potential effects of climate change on model results. Regional statistics differ substantially among neighboring rain zones (table 2) and among neighboring ecoregions (Granato, 2010); they do not necessarily characterize precipitation statistics for any particular drainage basin within each zone (or region), but are intended to be the basis for producing initial planning-level estimates for a typical basin within that zone (or region).

Table 2. Summary of synoptic storm-event precipitation statistics for the 15 U.S. Environmental Protection Agency rain zones within the conterminous United States.

[The statistics are calculated from data collected at selected hourly-precipitation data stations in each rain zone during the 1965–2006 period (Granato, 2010). COV, coefficient of variation is defined as the standard deviation divided by the mean; storm events are defined as having a minimum volume of 0.1 inches of precipitation and a 6-hour minimum interevent time threshold. Rain zones are shown on figure 6.]

| Rain zone | | Number of stations | Number of storm events (per year) | | Annual precipitation volume (inches) | | Storm-event volume (inches) | | Storm-event duration (hours) | | Time between storm-event midpoints (hours) | |
Number	Name		Mean	COV	Mean	COV	Mean	COV	Mean	COV	Mean	COV
1	Northeast	270	55	0.29	30.81	0.30	0.56	0.93	8.4	0.85	148.7	1.19
2	Northeast Coastal	61	53	0.26	37.59	0.30	0.71	1.00	9.3	0.81	153.9	1.04
3	Mid-Atlantic	122	49	0.29	32.67	0.31	0.67	0.99	7.8	0.88	166.2	1.26
4	Central	182	50	0.29	35.57	0.31	0.71	1.00	7.2	0.90	164.5	1.30
5	North Central	506	42	0.30	24.70	0.32	0.59	0.97	7.0	0.88	210.5	1.46
6	Southeast	227	53	0.25	43.19	0.28	0.82	1.04	6.8	0.93	158.8	1.25
7	East Gulf	84	57	0.23	47.56	0.27	0.84	1.11	5.7	1.00	150.1	1.36
8	East Texas	173	37	0.26	28.79	0.30	0.78	1.07	6.5	0.96	239.3	1.34
9	West Texas	63	24	0.30	14.95	0.35	0.62	1.00	5.6	1.00	375.4	1.72
10	Southwest	157	23	0.34	10.15	0.38	0.45	0.87	6.2	0.95	427.7	1.64
11	West Inland	72	17	0.43	11.48	0.51	0.60	0.99	8.6	0.91	675.7	1.70
12	Pacific Southwest	89	17	0.42	14.90	0.54	0.85	1.13	10.4	0.94	498.9	2.07
13	Northwest Inland	353	27	0.34	11.48	0.36	0.42	0.87	7.4	0.91	356.6	1.62
14	Pacific Central	134	35	0.34	29.97	0.41	0.84	1.10	12.0	0.92	255.1	2.05
15	Pacific Northwest	117	62	0.26	46.03	0.31	0.70	1.11	12.6	0.93	159.4	1.74

Prestorm Streamflow Volumes

SELDM is designed to calculate prestorm streamflow volumes from the basin upstream of the highway-runoff mixing point for each storm as a stochastic variable. Prestorm streamflow is one component of the total stormflow from the upstream basin (fig. 5). Granato (2010) provides a detailed discussion of the methods and data used for estimating prestorm streamflows for use with SELDM. The prestorm streamflow, modeled as the instantaneous flow at the beginning of a storm, is added to the current storm runoff for the duration of the highway runoff or BMP discharge to model the total flow available for dilution in the current storm. In the environment, the components of prestorm flow may include base flow (generally defined as groundwater discharge) and stormflow from a previous storm. In the environment and in SELDM analyses, some proportion of prestorm flows may equal 0 if the stream is intermittent or ephemeral. Estimates of prestorm streamflow in receiving waters are important for assessing risks for adverse effects of runoff on water quality, because prestorm flow can be a substantial proportion of total stormflow. The prestorm-streamflow statistics available in SELDM were calculated by using data from the 2,783 selected U.S. Geological Survey streamgages in the conterminous United States shown in figure 8 (Granato, 2010).

Granato (2010) demonstrated that the population of prestorm flows is well represented by the complete population of daily mean streamflows. This concept is shown schematically in figure 9. This schematic diagram shows that the range of prestorm flows might be wide if the definition of the minimum time between precipitation events is less than the stormflow-recession duration. These patterns are apparent in daily mean flow data from many streams in the United States. Approved streamflow data are reported as daily mean flows by the USGS (Mathey, 1998; U.S. Geological Survey, 2011). In comparison, independent storm events are commonly defined by using hourly data and by specifying an interevent time, which is the minimum number of dry hours between independent storm events (Driscoll and others, 1979; Athayde and others, 1983; Adams and others, 1986; Driscoll, Palhegyi, and others, 1989; U.S. Environmental Protection Agency, 1992; Wanielista and Yousef, 1993; Guo and Adams, 1998a; Adams and Papa, 2000; Granato, 2010). The minimum interevent time may differ considerably among regions but is generally approximated by an interval of about 6 hours (Driscoll, Palhegyi, and others, 1989; Granato, 2010). Theoretically, there may be as many as four independent storm events with an event duration of 1 hour and a minimum interevent time of 6 hours during one 24-hour period used for reporting one daily mean streamflow value. Runoff events commonly are defined by the duration of the stormflow hydrograph (Linsley and others, 1975; Chow and others, 1988). Prestorm flows may include runoff from a previous storm because stormflow-recession durations for many basins commonly are longer than one or more days (Linsley and others, 1975; Sloto and Crouse, 1996). Despite the difference between the operational definitions of storm events and runoff events, daily mean streamflow statistics commonly are used as an approximation for receiving-water flow during storm events (Di Toro, 1984; Driscoll, Shelley, and others, 1989; Driscoll and others, 1990a, b; Novotny, 2004). Granato (2010), however, demonstrated that daily mean streamflow statistics may be a better approximation for representing prestorm flow than stormflow because continuous-flow records commonly include a substantial proportion of dry days.

SELDM uses conditional-probability methods to account for the occurrence of prestorm flows equal to 0 and the log-Pearson Type III distribution to stochastically generate the remaining population of nonzero prestorm flows (table 1; appendix 1). A uniform random number between 0 and 1 is generated to represent the total probability (plotting position) of the zero and nonzero prestorm flow for each storm event. If this number is less than or equal to the proportion of zero flows, then a prestorm streamflow value of 0 is assigned for that storm event. If the initial uniform random number is greater than the proportion of zero flows, then this number is rescaled to the range of 0 to 1 to generate a frequency factor that represents the prestorm flow within the probability distribution of the nonzero streamflows.

Although the lognormal distribution is most commonly used for highway- and urban-runoff studies, a log-Pearson type III distribution was selected for generating stochastic planning-level estimates of prestorm streamflow by means of the frequency-factor method (equation 1). The log-Pearson type III distribution was selected because it is an extremely flexible distribution that can assume different shapes such as symmetrical, positively skewed, or negatively skewed (Haan, 1977; Chow and others, 1988; Bobee and Ashkar, 1991; Stedinger and others, 1993; Cheng and others, 2007; Granato, 2010). The log-Pearson Type III distribution is equivalent to the lognormal distribution if the logarithms of streamflow have zero skew. The log-Pearson Type III was selected because Granato (2010) found that coefficients of skew of the logarithms of nonzero flow measured at the 2,783 selected streamgages ranged from -2.2 to 5.4, and only 9 percent of the distributions had log-skew values within the 95-percent confidence limit of a lognormal distribution. The mean and standard deviation of the logarithms of nonzero streamflow data were used to calculate the location and spread of the prestorm flow values (equation 1). The skew coefficient was used to adjust the standard normal variates to produce a representative sample of data.

SELDM offers five options for selecting prestorm-flow statistics on the streamflow-statistics form. These options are selecting (1) the mean of streamgage statistics by ecoregion, (2) the median of streamgage statistics by ecoregion, (3) the mean of statistics from user-selected streamgages, (4) the median of statistics from user-selected streamgages, or (5) entering user-defined statistics. Methods for selecting these options on the statistics form are described in detail in appendix 4. The ecoregion is automatically selected by entering the latitude and longitude of the highway site. The

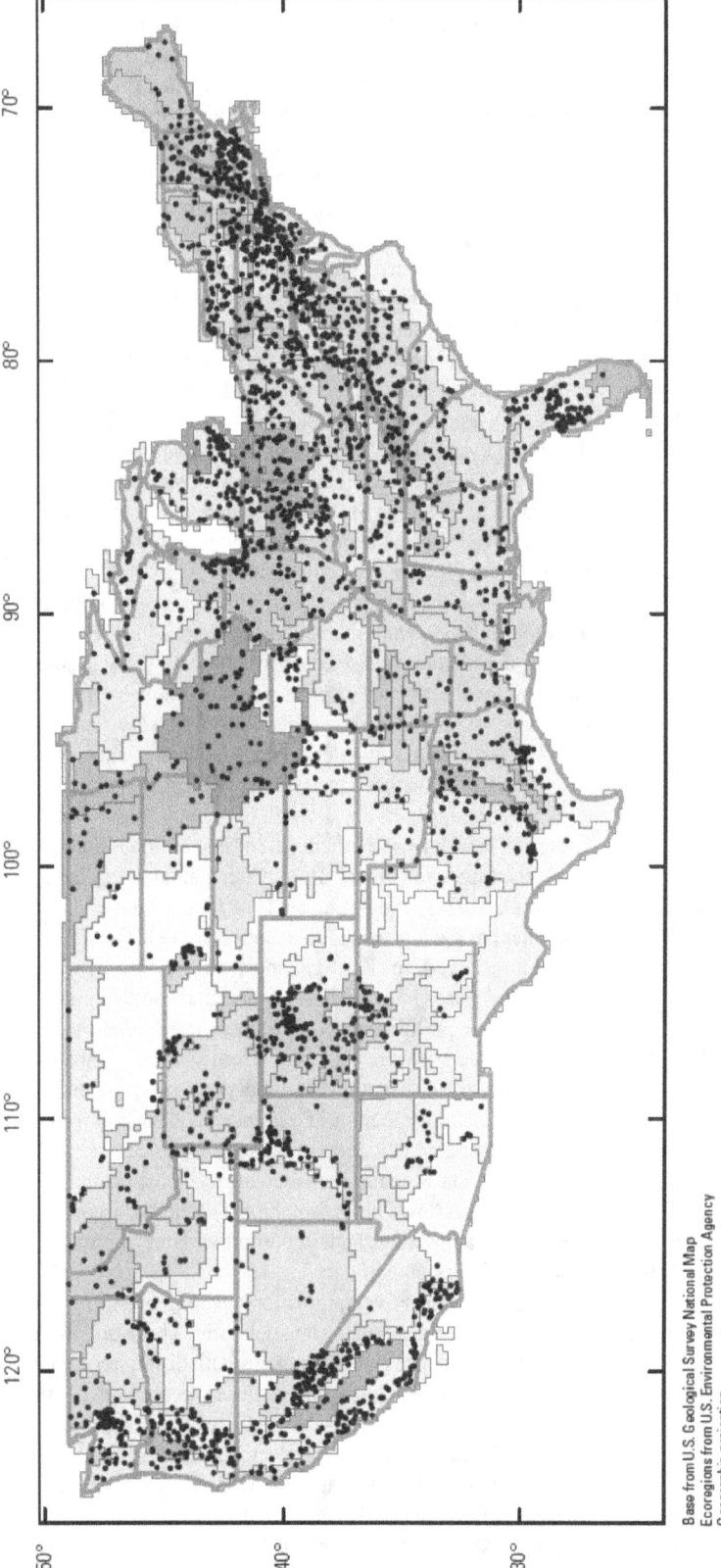

Base from U.S. Geological Survey National Map
Ecoregions from U.S. Environmental Protection Agency
Geographic projection

Figure 8. The spatial distribution of 2,783 selected U.S. Geological Survey streamgages (black dots) with respect to U.S. Environmental Protection Agency (2003) Level III ecoregions (colored polygons), which have been discretized to a 15-minute grid in the conterminous United States. Ecoregions are identified on the plate useco.pdf on the CD–ROM accompanying this report.

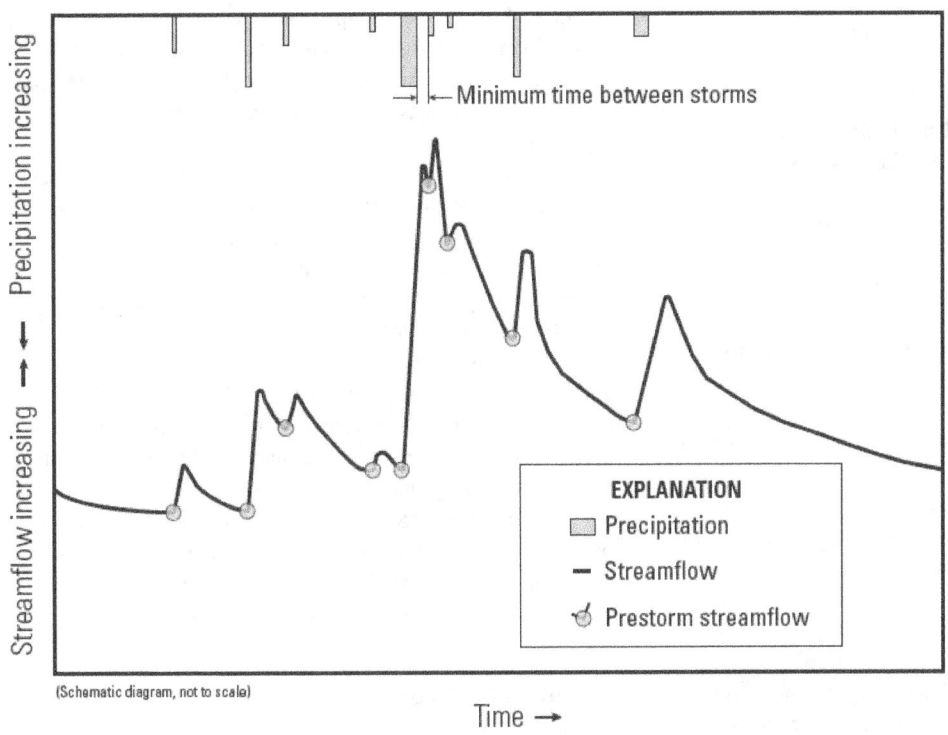

Figure 9. The potential variability in prestorm flows that may occur if the definition of the minimum time between precipitation events is less than the duration of stormflow recession for a given basin. The minimum time between storms for highway- and urban-runoff studies is 6 hours without measurable precipitation (Driscoll, Palhegyi, and others, 1989), whereas the stormflow recession for many basins may last longer than one or more days (Linsley and others, 1975).

user, however, can manually select an ecoregion that better represents conditions at a site of interest. The options for calculating prestorm flow statistics for the site of interest from the mean or median of statistics from selected streamgages are provided so that values from nearby hydrologically similar basins can be used. The option for entering user-defined statistics can be used to enter site specific statistics, to do a sensitivity analysis, or to evaluate potential effects of climate change on model results.

The options for calculating statistics for the site of interest from the mean or median of statistics for every site in an ecoregion are provided to formulate initial planning-level estimates. For example, Granato (2010) selected USEPA Level III ecoregions to provide regional planning-level estimates because ecoregions represent areas of hydrologic similarity on a large-scale basis. He showed ecoregions can be an effective tool for characterizing streamflows in different areas of the country. Also, ecoregions are commonly being used as a spatial framework for organizing and interpreting environmental data (Intergovernmental Task Force on Monitoring Water Quality, 1995a, b; Omernik and others,

2000; Omernik, 2004; McMahon and others, 2001; U.S. Environmental Protection Agency, 2003; Brown, 2006). Regional statistics differ substantially among neighboring ecoregions (table 3); they do not necessarily characterize prestorm-flow statistics for any particular drainage basin within each ecoregion, but are intended to be the basis for producing initial planning-level estimates for a typical basin within that ecoregion (Granato, 2010).

If regional statistics or statistics from nearby hydrologically similar sites are used, then SELDM uses the drainage-area-ratio method to calculate streamflow statistics for the site of interest (Granato, 2010). The assumption of hydrologic similarity is implicit in the application of the drainage-area-ratio method because basin characteristics are not explicitly included in the predictive equation. Natural factors (such as orographic effects, variations in soils, and geology) and anthropogenic factors (such as total impervious fraction and water use) should be considered in assessments of hydrologic similarity. The general equation for the drainage-area-ratio method is

$$Q_y = \left(A_y / A_x \right)^Z \times Q_x, \qquad (4)$$

where

Q_y is the estimated streamflow at the site of interest in cubic feet per second (ft³/s),

Q_x is the streamflow at the index site in ft³/s,

A_y is the drainage area for the site of interest in square miles (mi²),

A_x is the drainage area for the index site in mi², and

Z is the streamflow exponent, commonly assumed to equal 1.

The default application of the drainage-area-ratio method for generating prestorm flows for the site of interest is based on the assumption that the exponent Z is 1. Granato (2010) found that this was not an unreasonable assumption for most ecoregions and provided regression equations that adjust flow statistics for each ecoregion to account for the effects of divergent Z values. SELDM uses the drainage area of the site of interest multiplied by the normalized geometric mean streamflow for the ecoregion (Q_x/A_x), in cubic feet per second per square mile (ft³/s/mi²) for selected sites to estimate the necessary mean values. The standard deviation and skew of the logarithms of nonzero flows are unitless and therefore do not need to be adjusted by drainage area under the assumption that Z is 1. If the exponent Z varies substantially from 1, then statistics from hydrologically similar sites with similar drainage areas may be used to estimate the flow statistics at the site of interest. In this case, the flow statistics may be estimated by select nearby sites in SELDM or by using the methods described by Granato (2010) to calculate statistics based on data from other nearby sites. Alternatively, the regression equations provided by Granato (2010) for predicting the mean, standard deviation, and skew of the logarithms of nonzero flows from drainage areas may be used to calculate improved regional estimates. The results may be entered as user-defined values on the streamflow-statistics form.

Granato (2009) developed five programs for obtaining and analyzing streamflow data from the USGS National Water Information System Web site (NWISWeb) in support of the SELDM development project. The Get National Water Information System Streamflow (Q) files (GNWISQ) program was written for obtaining daily mean streamflow data. The Streamflow (Q) Statistics (QSTATS) program was written to calculate the statistics used by SELDM. The Streamflow Record Extension Facilitator (SREF) program was written to provide a long-term record of daily mean streamflows (record extension) or estimates of long-term streamflow statistics (record augmentation) for sites with limited data. The Make Plotting Position file (MkPP) was written to facilitate generation of flow-duration curves for assessing hydrologic similarity in the flow record among potential index streamflow-monitoring sites. The Make U.S. Environmental

Agency DFLOW3 batch input Files (MkDFlowF) program was written to facilitate batch use of DFLOW3 to calculate low-flow statistics for many streamflow records; such statistics can be used as a measure of hydrologic similarity among monitoring sites.

Granato (2010) used streamflow data from 2,783 long-term streamgages to provide prestorm-flow estimates, but additional data are available for more detailed regional or local studies. Continuous records of daily mean streamflow for periods of years to decades are available for almost 26,000 streamgages across the United States. The USGS (2011) indicates that concurrent measurements of stage and flow are available for almost 53,000 sites. Data from one or more paired measurements of stream discharge and concentration measured at more than 43,000 surface-water-quality monitoring stations in the conterminous United States are available in NWISWeb (Granato and others, 2009). The programs developed by Granato (2009) may be useful for establishing new streamflow datasets for the SELDM database or providing user-defined statistics for a given site of interest.

Storm-Runoff Volumes

SELDM is designed to calculate runoff volumes for each storm as a stochastic variable (appendix 1). Runoff coefficients are used with the randomly generated precipitation volumes to calculate the volume of runoff from the highway site and the upstream basin for each storm (fig. 5). A truncated Pearson type III distribution was selected for generating stochastic planning-level estimates of runoff coefficients for the highway site and the upstream basin by the frequency-factor method (equation 1) (Haan, 1977; Interagency Advisory Committee on Water Data, 1982; Chow and others, 1988; Stedinger and others, 1993; Cheng and others, 2007; Granato, 2010). Runoff coefficient values from a Pearson Type III distribution are estimated using the frequency-factor method (equation 1) because it is an extremely flexible distribution that can assume different shapes such as symmetrical, positively skewed, or negatively skewed (Haan, 1977; Chow and others, 1988; Bobee and Ashkar, 1991; Granato, 2010). The mean and standard deviation of runoff-coefficient data are used to calculate the location and spread of the resultant runoff coefficients (equation 1). The skew coefficient is used to adjust the standard normal variates to produce a representative sample of data. If the skew of a population equals 0, the frequency factor is the standard normal variate. As skew coefficients deviate from 0, the relation between the plotting-position probability and the associated frequency factor shifts to reflect the distribution of values above and below the median value. Although the Pearson type III distribution is not bounded by 0 and 1, modeled runoff coefficients are limited to this range because the model generates each storm as an independent event. Granato (2010) discusses hydrologic conditions and sources of uncertainty in rainfall-runoff data that may result in runoff coefficients larger than 1.

Table 3. Medians of selected streamflow statistics for the 84 U.S. Environmental Protection Agency Level III nutrient ecoregions in the conterminous United States calculated by using daily mean streamflow data from 2,783 selected U.S. Geological Survey streamgages for the period 1960–2003. Statistics include the proportion of zero flows and the mean, standard deviation, and skew of the logarithms of nonzero mean daily streamflow measurements.

[No., number; ft^3/s/mi^2, cubic foot per second per square mile; mi^2, square mile; SD, standard deviation. Ecoregions are identified on the plate useco.pdf on the CD–ROM accompanying this report]

Ecoregion		Number of sta-tions	Days with zero discharge (percent of total record)	Statistics for the common logarithms of nonzero discharges		
No.	Name			Geometric mean (ft^3/s/mi^2)	Geometric SD (dimensionless)	Coefficient of skew (dimensionless)
1	Coast Range	35	0.00	1.77	3.99	0.11
2	Puget Lowland	28	0.00	2.14	2.37	0.26
3	Willamette Valley	15	0.00	1.21	4.22	-0.03
4	Cascades	72	0.00	2.72	2.59	0.06
5	Sierra Nevada	86	0.00	0.50	3.92	0.41
6	Southern and Central California Plains and Hills	113	5.46	0.08	7.18	0.19
7	Central California Valley	8	40.59	0.08	7.46	-0.15
8	Southern California Mountains	17	15.32	0.04	7.20	0.02
9	Eastern Cascades Slopes and Foothills	11	0.00	1.67	2.25	0.43
10	Columbia Plateau	17	0.00	0.22	4.90	0.06
11	Blue Mountains	15	0.00	0.64	2.91	0.50
12	Snake River Basin/High Desert	3	45.56	0.06	5.00	-0.19
13	Northern Basin and Range	30	0.00	0.23	2.75	0.75
14	Southern Basin and Range	6	0.02	0.07	4.50	0.20
15	Northern Rockies	11	0.00	0.72	3.02	0.58
16	Montana Valley and Foothill Prairies	14	0.00	0.46	2.76	0.91
17	Middle Rockies	48	0.00	0.39	2.91	0.63
18	Wyoming Basin	29	0.00	0.28	3.61	0.70
19	Wasatch and Uinta Mountains	40	0.00	0.37	2.83	0.87
20	Colorado Plateaus	25	0.00	0.20	2.75	0.70
21	Southern Rockies	114	0.00	0.39	3.13	0.72
22	Arizona/New Mexico Plateau	22	0.00	0.13	2.71	0.64
23	Arizona/New Mexico Mountains	19	0.00	0.08	2.92	0.82
24	Southern Deserts	5	98.32	0.03	18.40	0.08
25	Western High Plains	15	0.00	0.10	2.49	0.37
26	Southwestern Tablelands	14	0.00	0.03	3.32	0.25
27	Central Great Plains	35	0.97	0.03	5.34	0.18
28	Flint Hills	6	0.40	0.09	5.60	0.07
29	Central Oklahoma/Texas Plains	31	10.20	0.04	8.06	0.31
30	Edwards Plateau	13	2.39	0.14	4.86	-0.31

Table 3. Medians of selected streamflow statistics for the 84 U.S. Environmental Protection Agency Level III nutrient ecoregions in the conterminous United States calculated by using daily mean streamflow data from 2,783 selected U.S. Geological Survey streamgages for the period 1960–2003. Statistics include the proportion of zero flows and the mean, standard deviation, and skew of the logarithms of nonzero mean daily streamflow measurements.—Continued

[No , number; ft³/s/mi², cubic foot per second per square mile; mi², square mile; SD, standard deviation. Ecoregions are identified on the plate useco.pdf on the CD–ROM accompanying this report]

Ecoregion		Number of sta- tions	Median of streamflow statistics for each ecoregion			
			Days with zero discharge (percent of total record)	Statistics for the common logarithms of nonzero discharges		
No.	Name			Geometric mean (ft³/s/mi²)	Geometric SD (dimensionless)	Coefficient of skew (dimensionless)
31	Southern Texas Plains	7	58.27	0.01	8.29	0.48
32	Texas Blackland Prairies	28	8.77	0.06	9.07	0.10
33	East Central Texas Plains	11	5.90	0.05	8.26	0.00
34	Western Gulf Coastal Plains	29	0.00	0.20	5.08	0.56
35	South Central Plains	40	0.22	0.23	6.75	-0.01
36	Ouachita Mountains	7	2.67	0.23	9.16	-0.29
37	Arkansas Valley	4	2.82	0.20	10.17	-0.54
38	Boston Mountains	6	1.65	0.30	8.20	-0.65
39	Ozark Highlands	24	0.00	0.43	3.39	0.35
40	Central Irregular Plains	38	0.58	0.10	7.52	0.06
41	Canadian Rockies	3	0.02	1.30	3.57	0.07
42	Northwestern Glaciated Plains	12	0.35	0.21	3.38	0.16
43	Northwestern Great Plains	27	0.02	0.12	3.05	0.68
44	Nebraska Sandhills	3	0.00	0.24	1.75	1.76
45	Piedmont	112	0.00	0.66	2.73	0.20
46	Northern Glaciated Plains	22	32.98	0.02	10.81	0.13
47	Western Corn Belt Plains	56	0.02	0.19	4.27	0.01
48	Lake Agassiz Plain	12	3.96	0.03	6.19	0.07
49	Northern Minnesota Wetlands	1	1.25	0.05	9.19	-0.38
50	Northern Lakes and Forests	45	0.00	0.69	2.13	0.74
51	Northern Central Hardwood Forests	14	0.00	0.34	2.65	0.60
52	Driftless Area	13	0.00	0.49	1.88	1.27
53	Southeastern Wisconsin Till Plains	16	0.00	0.39	2.84	0.36
54	Central Corn Belt Plains	71	0.00	0.39	3.65	0.04
55	Eastern Corn Belt Plains	87	0.00	0.34	4.25	0.24
56	S. Michigan/N. Indiana Drift Plains	70	0.00	0.53	2.41	0.13
57	Huron/Erie Lake Plains	11	0.00	0.25	4.13	0.36
58	Northeastern Highlands	107	0.00	1.09	2.90	0.09
59	Northeastern Coastal Zone	79	0.00	1.02	2.90	-0.16
60	Northern Appalachian Plateau and Uplands	31	0.00	0.68	3.20	0.11

Table 3. Medians of selected streamflow statistics for the 84 U.S. Environmental Protection Agency Level III nutrient ecoregions in the conterminous United States calculated by using daily mean streamflow data from 2,783 selected U.S. Geological Survey streamgages for the period 1960–2003. Statistics include the proportion of zero flows and the mean, standard deviation, and skew of the logarithms of nonzero mean daily streamflow measurements.—Continued

[No., number; ft³/s/mi², cubic foot per second per square mile; mi², square mile; SD, standard deviation. Ecoregions are identified on the plate useco.pdf on the CD–ROM accompanying this report]

Ecoregion		Number of sta-tions	Median of streamflow statistics for each ecoregion			
			Days with zero discharge (percent of total record)	Statistics for the common logarithms of nonzero discharges		
No.	Name			Geometric mean (ft³/s/mi²)	Geometric SD (dimensionless)	Coefficient of skew (dimensionless)
61	Erie/Ontario Lake Hills and Plain	21	0.00	0.65	3.20	0.19
62	North Central Appalachians	35	0.00	1.03	2.98	-0.07
63	Middle Atlantic Coastal Plain	23	0.00	0.51	3.85	-0.00
64	Northern Piedmont	94	0.00	0.78	2.58	0.31
65	Southeastern Plains	100	0.00	0.64	3.04	0.28
66	Blue Ridge Mountains	35	0.00	1.84	1.99	0.36
67	Central Appalachian Ridges and Valleys	105	0.00	0.75	2.73	0.39
68	Southwestern Appalachians	15	0.00	0.67	4.87	0.02
69	Central Appalachians	54	0.00	0.78	3.63	-0.05
70	Western Allegheny Plateau	54	0.00	0.54	4.08	-0.18
71	Interior Plateau	59	0.00	0.48	4.70	0.02
72	Interior River Lowland	33	0.55	0.16	7.72	0.02
73	Mississippi Alluvial Plain	7	0.00	0.64	4.32	0.09
74	Mississippi Valley Loess Plains	7	0.00	0.64	2.73	1.52
75	Southern Coastal Plain	93	0.20	0.29	4.55	-0.20
76	Southern Florida Coastal Plain	1	58.48	2.40	4.43	-1.44
77	North Cascades	15	0.00	4.67	2.44	0.11
78	Klamath Mountains	34	0.00	0.80	3.68	0.22
79	Madrean Archipelago	3	3.57	0.004	3.62	0.32
80	Northern Basin and Range	13	0.00	0.10	3.37	0.40
81	Sonoran Basin and Range	11	27.52	0.08	8.10	-0.11
82	Laurentian Plains and Hills	11	0.00	1.07	2.86	0.16
83	Eastern Great Lakes and Hudson Lowlands	43	0.00	0.71	3.04	0.18
84	Atlantic Coastal Pine Barrens	34	0.00	1.04	1.96	0.11

Regression equations relating the mean, standard deviation, and skew of runoff coefficients to the total impervious fraction were developed to facilitate the selection of representative statistics for the highway site and the upstream basin (fig. 10). Regression equations that are based on Schueler's (1987) analysis of NURP data also are available for use in SELDM, and user-defined runoff-coefficient statistics also may be used. The regression equations for highway sites were developed on the basis of data from 58 highway sites across the country with 9 or more storm events. The drainage areas of these sites range from 0.05 to 106 acres and TIA fractions from 0.27 to 1. Regression equations for estimating runoff-coefficient statistics for the upstream basin were developed on the basis of data from 167 sites across the country with 9 or more storm events (Granato, 2010). The drainage areas of these sites range from 0.005 to 93.47 mi^2 and TIA fractions from 0.0001 to 0.994. Separate regression equations were developed for the two sets of sites because highway sites commonly are smaller and more homogenous than the other sites. If the impervious fraction of the highway site (or upstream basin) is less than about 0.3, then the regression equations developed with the 167 nonhighway monitoring sites will provide more representative statistics than the regression equations developed by using only highway sites (fig. 10).

The mean of the runoff coefficients for the highway and the upstream basin can be estimated with respect to the impervious fraction (*IFs*) with the regression equations developed for use with SELDM. A one-segment regression model was developed by using the highway-site data because no break in slope was apparent in the mean runoff coefficient values for highway sites (fig. 10A). This equation is

$$R_vMean = 0.03 + 0.755 \times IF . \tag{5}$$

In this equation, the runoff coefficients (*R_vMean*) and the *IF* are unitless and range from 0 to 1. The slope of this regression line is significantly different from 0 within a 95-percent confidence interval. A two-segment regression model was developed by using data from the nonhighway sites to account for a change in slope of the relation between *R_vMean* and *IF*. The breakpoint between the best-fit segments is at an *IF* value of 0.55. These equations are

$$R_vMean = 0.129 + 0.225 \times IF \tag{6}$$

if the *IF* value is less than or equal to 0.55 and

$$R_vMean = -0.371 + 1.14 \times IF \tag{7}$$

if the *IF* is greater than or equal to 0.55 (fig. 10A). The slopes of both segments are significantly different from 0 within a 95-percent confidence interval. The two-segment regression model accounts for the steeper trend in site-mean runoff coefficients above an *IF* value of about 0.55. The equations developed for the highway sites and the upstream basins

both produce mean runoff coefficients that are less than 0.8 for sites that are completely impervious. A site-mean runoff coefficient of 0.8 may seem low for a completely impervious area, but studies show that evaporation and infiltration from paved surfaces may have mean values in the range from about 20 to more than 30 percent over many storms (Mansell and Rollet, 2006; Ramier and others, 2006; Wiles and Sharp, 2008; Wanielista and others, 2010).

The standard deviations of the runoff coefficients (*R_vSD*) for the highway and the upstream basin (fig. 10B) also are unitless. One-segment regression models were developed from the highway-site data and data from the nonhighway sites because no breaks in slope were apparent in the *R_vSD* values for both types of site. The highway-site equation is

$$R_vSD = 0.229 - 0.0373 \times IF , \tag{8}$$

and the upstream-basin equation is

$$R_vSD = 0.099 + 0.015 \times IF . \tag{9}$$

Neither of these slopes is significantly different from 0 within a 95-percent confidence interval. For the standard deviation, use of the median value, which is about 0.21 for highway sites and about 0.10 for nonhighway sites, has about the same predictive power as the associated regression equation.

The coefficients of skew of the runoff coefficients (*R_vSk*) for the highway and the upstream basin (fig. 10C) also are unitless. A one-segment regression model was developed for the highway-site data because no break in slope was apparent:

$$R_vSkew = 2.13 - 3.32 \times IF . \tag{10}$$

The slope of this regression line is significantly different from 0 within a 95-percent confidence interval. A two-segment regression model was developed for the skew of runoff coefficients at nonhighway sites from the estimated *IF*. These equations are

$$R_vSkew = 1.08 - 0.557 \times IF \tag{11}$$

if the *IF* value is less than or equal to 0.52 and

$$R_vSkew = 2.22 - 2.73 \times IF \tag{12}$$

if the *IF* is greater than or equal to 0.52 (fig. 10C). The model accounts for the steeper trend in the coefficient of skew of runoff coefficients above an *IF* of about 0.52. The slope of the first segment is not significantly different from 0, but the slope of the second segment is significantly different from 0 within a 95-percent confidence interval. These slopes indicate that the skew coefficients vary randomly below an *IF* of 0.52 and generally decrease with increasing *IF* above this threshold (Granato, 2010).

The third predefined set of regression equations for calculating *R_v* statistics in SELDM is labeled "Schueler

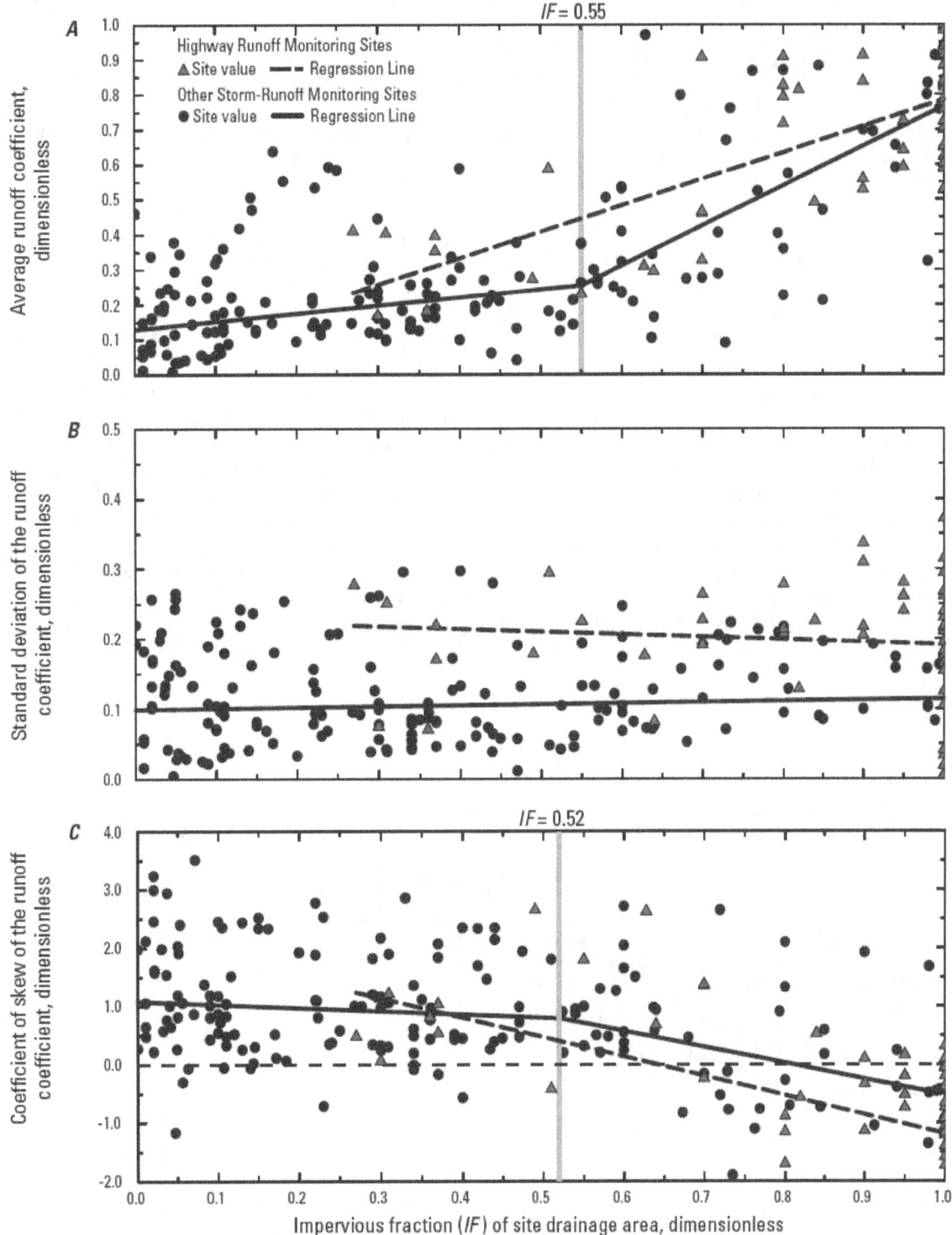

Figure 10. *A,* The mean, *B,* standard deviation, and *C,* coefficient of skew of runoff coefficients for 58 highway-runoff monitoring sites and 167 other storm-runoff monitoring sites with 9 or more storm events. Nonparametric regression lines indicate the relation between each statistic and the impervious fraction *(IF)* in the respective drainage areas.

Trimmed NURP." This set of equations is based on Schueler's (1987) selection of 43 sites with impervious fractions ranging from 0.01 to 1.0 from the USEPA NURP report (Athayde and others, 1983). The equations are provided for comparison with results from other studies or with average calculations developed by using Schueler's "Simple Method" (Schueler, 1987). The regression equation for the mean R_v value is Schueler's (1987) equation. The regression equation for the standard deviation was calculated from the coefficient of variation (COV, which is the standard deviation divided by the mean) of each of these sites reported by Athayde and others (1983). Athayde and others (1983) indicated that the R_v values for individual NURP sites were lognormally distributed; therefore, the skew coefficients were estimated from the COV by using a theoretical equation (Stedinger and others, 1993). The regression equation was developed to estimate the skew of R_v values with respect to the impervious fraction based on these theoretical skew values.

SELDM can model effects of antecedent conditions on the coefficients for runoff from the upstream basin by using the prestorm streamflow as the explanatory variable. If a nonzero correlation coefficient is entered on the runoff-coefficient statistics form, SELDM uses the methods described for generating correlated random numbers to calculate the plotting position of the runoff coefficients from the plotting positions of the prestorm flows. Granato (2010) calculated

rank correlation coefficients to evaluate potential relations between prestorm streamflow and runoff coefficients. Although prestorm streamflow is used as the explanatory variable, correlation does not necessarily imply causation. For example, antecedent precipitation may saturate soils and increase prestorm streamflow. In this case, higher prestorm streamflow may indicate wetter antecedent conditions, but not necessarily cause more runoff.

In the storm-event database compiled for the SELDM study, 43 sites have at least 7 paired prestorm-streamflow and runoff-coefficient values (Granato, 2010). Figure 11 shows the Spearman's rho values for these datasets and the associated 95-percent confidence intervals, which are functions of sample size (Caruso and Cliff, 1997). Three sites had very weak negative correlations. Ten sites had positive rho values that were less than about 0.3, indicating that variations in prestorm streamflow may be associated with less than 30 percent of the variations in runoff coefficients from storm to storm at each of these sites. Seven sites have rho values between 0.3 and 0.5, indicating that prestorm streamflow may be associated with 30 to 50 percent of the variations in runoff coefficients at each of these sites. An additional 14 sites have rho values between 0.5 and 0.71, indicating that prestorm streamflow may be associated with 50 to 71 percent of variations in runoff coefficients at each of these sites. Nine sites have rho values that are greater than 0.71,

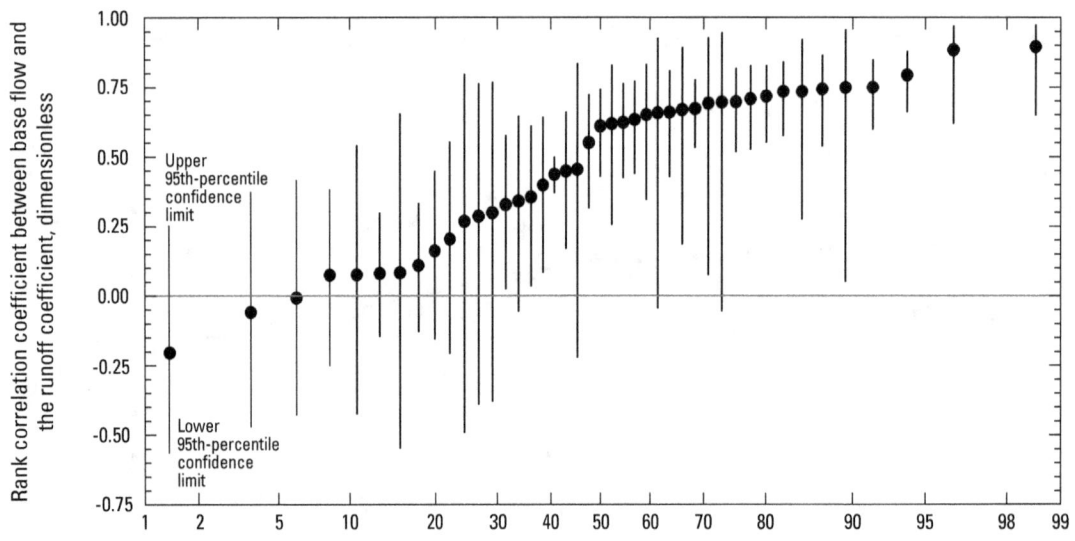

Figure 11.   The upper and lower 95th-percentile confidence limits and the mean values of the nonparametric rank correlation coefficients (Spearman's rho) for each of the 43 monitoring sites with 7 or more paired base-flow and runoff measurements (Granato, 2010).

indicating a moderate to strong correlation between prestorm streamflow and runoff coefficients at these sites. Differences in correlation coefficients among sites may reflect hydrologic-basin characteristics, artifacts in the assembled dataset, such as the use of different hydrograph-separation techniques in each study, or uncertainty in the samples (Granato, 2010). Seventeen sites had 95-percent confidence intervals that included 0; this result indicates that the true rho value may not be different from 0. Conversely, however, 35 of the sites have upper confidence limits that are greater than 0.5, which may indicate a substantial correlation between prestorm streamflow and the runoff coefficient. Rank correlations are user defined because prestorm streamflows vary substantially between different areas of the nation, and this dataset was too small to develop predictive relations for these rank correlation values and prestorm flows on the basis of available basin properties.

SELDM models potential relations between runoff coefficients for the upstream basin and the highway site by using the rank correlation between runoff coefficients because the potential for water-quality excursions depends, in part, on the volume of runoff from each source. The antecedent conditions that affect runoff from the upstream basin also may affect runoff from the highway site. Modeling the runoff coefficients without any correlation excludes the fact that the sites are adjacent and are subject to similar antecedent conditions, whereas modeling the runoff coefficients with perfect correlation excludes the fact that runoff-generating processes for the highway site and the upstream basin may be different, especially if the land use and drainage characteristics of these areas are different. An extensive literature search for information about correlation between runoff coefficients from adjacent sites was conducted, but no information was found. Four hydrologic experts were queried, but none of the respondents knew of any available literature on this topic (Dr. W.C. Huber, Oregon State University, written commun., September 2010; Dr. L.W. Mays, Arizona State University, written commun., September 2010; Dr. R.H. McCuen, University of Maryland, written commun., September 2010; Dr. M.P. Wanielista, Water Research Center, University of Central Florida, written commun. September, 2010).

An algorithm for calculating the rank correlation coefficient between runoff coefficients for the upstream basin and the highway site was developed under three general assumptions: (1) some degree of correlation would be expected because antecedent conditions and storm characteristics in nearby source areas would be similar; (2) the strength of the correlation would improve with increasing hydrologic similarity between the upstream and highway sites; and (3) the maximum correlation for basins with the same or similar impervious fractions would increase as the impervious fractions increase because runoff coefficients for different pervious areas are expected to be more variable than for different impervious areas for a given storm. Large differences in drainage areas between the upstream and highway sites are expected to introduce variability in the amount and distribution of precipitation. SELDM does not directly account

for spatial variation in precipitation because it is a lumped parameter model, but a reduced correlation between runoff coefficients may help to account for such effects. Large differences in total imperviousness between the highway site and the upstream basin may be a surrogate for large differences in drainage area because imperviousness tends to decrease with increasing drainage areas in stream basins (Granato, 2010). Comprehensive datasets are not available to confirm or quantify correlations based on these assumptions, but the hydrologic experts who responded to the queries for information agreed with these general assumptions (Dr. W.C. Huber, Oregon State University, written commun., September 2010; Dr. L.W. Mays, Arizona State University, written commun., September 2010; Dr. R.H. McCuen, University of Maryland, written commun., September 2010; Dr. M.P. Wanielista, Water Research Center, University of Central Florida, written commun., September 2010).

This algorithm was designed to use the absolute value of the difference in impervious fractions to calculate the rank correlation coefficient between runoff coefficients for the upstream basin and the highway site. Linear interpolation between impervious fractions and rank correlation coefficients was selected as a simple approximation in the absence of experimental data. The assumption of hydrologic similarity was implemented by calculating the correlation coefficient to generate a local maximum value for two sites with the same impervious fraction. The assumption that the correlation coefficient will increase as the impervious fractions of both the highway site and the upstream basin increase was implemented by making the local maximum and minimum (the floor) values functions of both the impervious fraction and the difference in impervious fractions. The linear functions for defining the maximum possible correlation coefficients (the ceilings) for the highway site and the upstream basin are

$$Rho_{CH} = RhoC_0 + \left(RhoC_1 - RhoC_0\right) \times IF_H, \quad (13)$$

and

$$Rho_{CU} = RhoC_0 + \left(RhoC_1 - RhoC_0\right) \times IF_U, \quad (14)$$

where

IF_H is the impervious fraction of the highway site,
IF_U is the impervious fraction of the upstream basin,
$RhoC_0$ is the ceiling when the impervious fraction equals 0,
$RhoC_1$ is the ceiling when the impervious fraction equals 1,
Rho_{CH} is the local value of the ceiling associated with the impervious fraction of the highway site, and
Rho_{CU} is the local value of the ceiling associated with the impervious fraction of the upstream-basin site.

Similarly, the linear functions for defining the minimum possible correlation coefficients (the floors) for the highway site and the upstream basin are

$$Rho_{FH} = RhoF_0 + \left(RhoF_1 - RhoF_0 \right) \times IF_H, \quad (15)$$

and

$$Rho_{FU} = RhoF_0 + \left(RhoF_1 - RhoF_0 \right) \times IF_U, \quad (16)$$

where

$RhoF_0$ is the floor when the impervious fraction equals 0,

$RhoF_1$ is the floor when the impervious fraction equals 1,

Rho_{FH} is the floor associated with the impervious fraction of the highway site, and

Rho_{FU} is the floor associated with the impervious fraction of the upstream-basin site.

The rank correlation value is then calculated by interpolating between the local ceiling and floor values according to the difference in impervious fractions and averaging the result for the highway and upstream basin:

$$Rho_{Rv} = \frac{\left(\left(Rho_{CH} - \left(Rho_{CH} - Rho_{FH} \right) \times |IF_H - IF_U| \right) + \left(Rho_{CU} - \left(Rho_{CU} - Rho_{FU} \right) \times |IF_H - IF_U| \right) \right)}{2}. \quad (17)$$

Implementation of this algorithm to generate Rho_{Rv} values that are greater than 0 requires that the input values are selected so that maximum distance between the ceiling ($RhoC$) and the floor ($RhoF$) will be less than the ceiling value for the whole range of impervious fractions. The ceiling and floor values were implemented as default values that can be changed in the table named "tblHighwayAnalysis" in the SELDM database for a given analysis or reset in the table design for all subsequent analyses. This default-value option was selected because neither available data nor the published hydrologic literature is sufficient to support either fixed values in the program code or guidance for user inputs in the model graphical user interface (GUI).

SELDM was implemented with ceiling and floor values that result in Spearman's rho values ranging from 0.9875 to 0.375 with local maxima where the impervious fractions are equal. The Spearman's rho values in figure 12 were calculated by using ceiling values of 0.5 and 0.9875 where the impervious fractions equal 0 and 1, and floor values of 0.25 and 0.5 where the impervious fractions equal 0 and 1. These values were selected to provide a high correlation between highly impervious and hydrologically similar sites and reduced but nonzero values as impervious fractions diverge. The minimum Rho_{Rv} value (0.375) corresponds to the maximum difference in impervious fractions (equation 17,

fig. 12) between completely impervious ($IF = 100$) and completely pervious ($IF = 0$) basins. Local Rho_{Rv} minimums along the X and Y axes coincide with local-maximum differences in impervious fractions along the axis. The values of Rho_{Rv} along the axes are equal to the means of the floor values (equation 17), which are functions of imperviousness (equations 15 and 16). Local maximums coincide with equal IF values; the Rho_{Rv} value equals the ceiling value for that impervious fraction, which in this case is a linear function of IF between 0.5 and 0.9875 (equations 13 and 14) The local maximums appear as the diagonal ridge on the contour plot (fig. 12).

SELDM generates a population of R_v values for the upstream basin and the highway site by using a four-step process: (1) the plotting positions for the R_v values of the upstream basin are generated by using the specified correlation to the plotting positions of upstream prestorm flow; (2) the upstream Rv value is calculated from upstream R_v statistics by using the frequency-factor method (equation 1) with Pearson Type III variates; (3) the plotting positions for the R_v values of the highway site are generated from the specified correlations to plotting positions for the R_v values of the upstream basin; and (4) the highway R_v value is calculated from highway R_v statistics by using the frequency-factor method (equation 1) with Pearson Type III variates. If the resultant R_v values are less than 0 or greater than 1, then the R_v values are set equal to 0 or equal to 1, respectively.

Storm-Event Hydrographs

SELDM is designed to calculate the event-mean concentrations and total storm loads for the entire highway-runoff event rather than intraevent flows and loads, but modeling the timing of flow from the highway, from the BMP outfall (if a BMP is selected), and from the upstream basin is necessary to calculate the potential amount of dilution during the period of discharge to the stream (fig. 13). Granato (2010) demonstrated that triangular runoff hydrographs commonly are used to model intraevent stormflows in hydraulic and water-quality models and are adequate for producing planning-level estimates for dilution analyses. It is necessary to model the intraevent stormflow hydrographs because differences in the locations, sizes, and drainage characteristics of the highway catchment and the upstream basin may cause differences in the timings and durations of runoff from each area. For example, if the highway catchment is small and the runoff drains directly to the stream, the duration of appreciable runoff from the highway catchment may be approximated

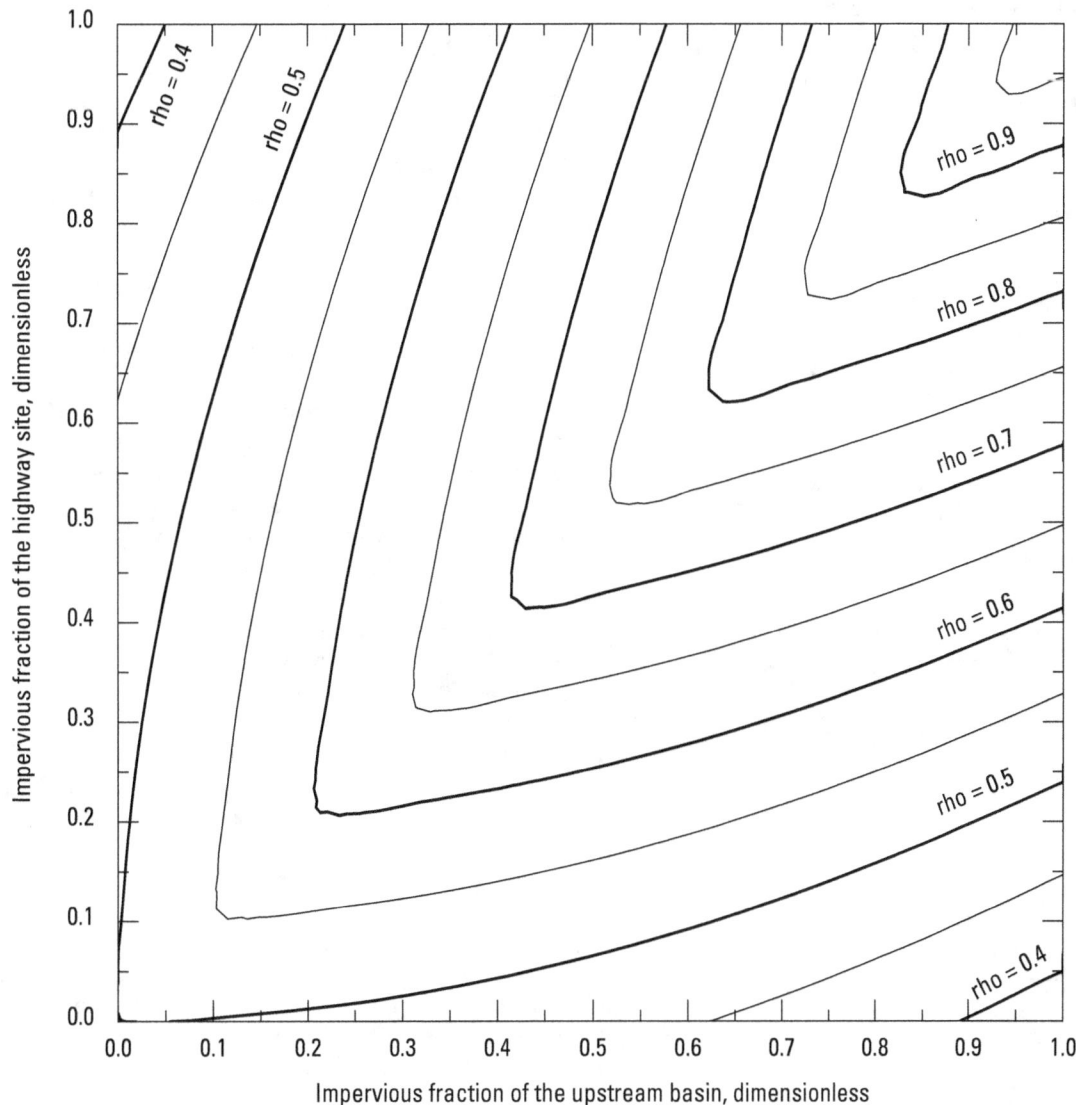

Figure 12.   Calculated Spearman's rho values showing the relations between the impervious fractions of the upstream basin and the highway site. These Spearman's rho values are calculated by using ceiling values of 0.5 and 0.9875 for impervious fraction values of 0 and 1 and floor values of 0.25 and 0.5 for impervious fraction values of 0 and 1, respectively.

A. Hypothetical triangular hydrographs

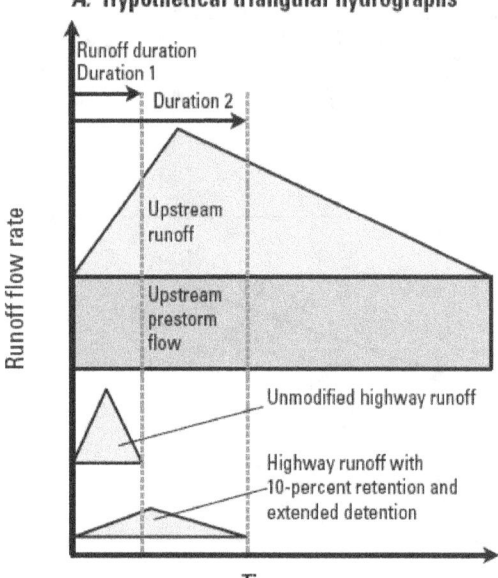

B. Hypothetical cumulative volumes

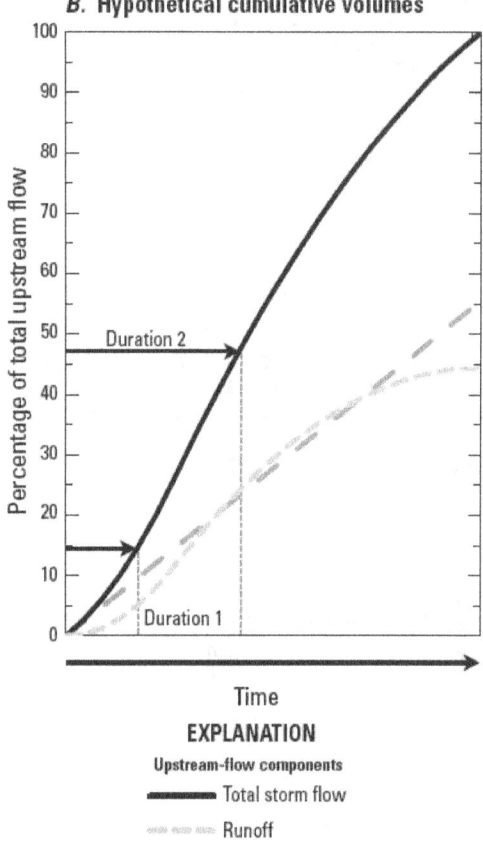

EXPLANATION

Upstream-flow components

———— Total storm flow

▪▪▪▪ Runoff

▬ ▬ Prestorm flow

Figure 13. *A*, Hypothetical triangular hydrographs, and *B*, the hypothetical cumulative upstream-stormflow volume that would coincide with unmodified runoff from a highway (duration 1) and with runoff from an extended detention structure (duration 2). This diagram shows the hypothetical runoff event with two upstream-flow components (runoff and prestorm flow), an unmodified highway-runoff hydrograph, and a highway-runoff hydrograph with retention and detention.

by the duration of the precipitation event. If the upstream basin is relatively large and more pervious than the highway catchment, appreciable runoff from the basin may continue for hours or days longer than runoff from the highway catchment. In this case, only a small proportion of the upstream runoff may be available to dilute highway-runoff constituents in the receiving waters. If, however, a structural BMP has been installed (or will be used) at the highway site to attenuate and extend the highway-runoff hydrograph, then a longer period of the upstream runoff may coincide with the period of highway runoff, and thus a larger volume of the upstream stormflow will be available to dilute highway-runoff constituents in the receiving waters.

This concept is demonstrated schematically in figure 13. In this hypothetical example, the triangular runoff hydrograph for the upstream basin is superimposed on a rectangular representation of the prestorm base-flow volume (fig. 13A). The durations of highway-runoff hydrographs with and without BMP modification are labeled "Duration 1" and "Duration 2." As indicated in the figure, a small increase in the duration of runoff from the highway may be accompanied by a large increase in the cumulative amount of concurrent runoff and base flow from the upstream basin, especially in the rising limb of the upstream-basin hydrograph (fig. 13B). If the highway-runoff or the BMP-discharge hydrograph extends beyond the end of the upstream-runoff hydrograph, then the available dilution equals the sum of the storm-event base flow (calculated as the product of the prestorm-flow rate and the duration of the storm), the storm-event runoff (fig. 13), and the poststorm base flow, which is calculated as the product of the prestorm-flow rate and the excess duration of the highway runoff or BMP discharge.

A triangular runoff hydrograph is used to calculate the amount of concurrent stormflow (fig. 13). The triangular runoff hydrograph can be fully parameterized with the total runoff volume, the start of runoff (T_o), the end of runoff (T_e), and the time to peak (T_p), respectively (fig. 14). Although the triangular hydrograph is a simple linear approximation, the CDF is an S-curve that is similar to the CDF for more complex approximations (Granato, 2010). Thus, the triangular hydrograph is sufficient for modeling the temporal distribution of flow during a storm. The proportion of total runoff at time T_i from the beginning of the storm for a triangular hydrograph is expressed as

$$R_c = \frac{(T_i - T_o)^2}{((T_e - T_o) \times (T_p - T_o))} \tag{18}$$

if $T_o <= T_i <= T_p$ and

$$R_c = 1 - \frac{(T_e - T_i)^2}{((T_e - T_o) \times (T_e - T_p))} \tag{19}$$

if $T_p <= T_i <= T_e$,

EXPLANATION

D Duration of rainfall-excess increment

BL Basin-lag time

T_p Time to peak (rising-limb time)

T_f Recession (falling-limb time)

T_b Hydrograph base time

T_o Begin time of the runoff hydrograph

T_e End time of the runoff hydrograph

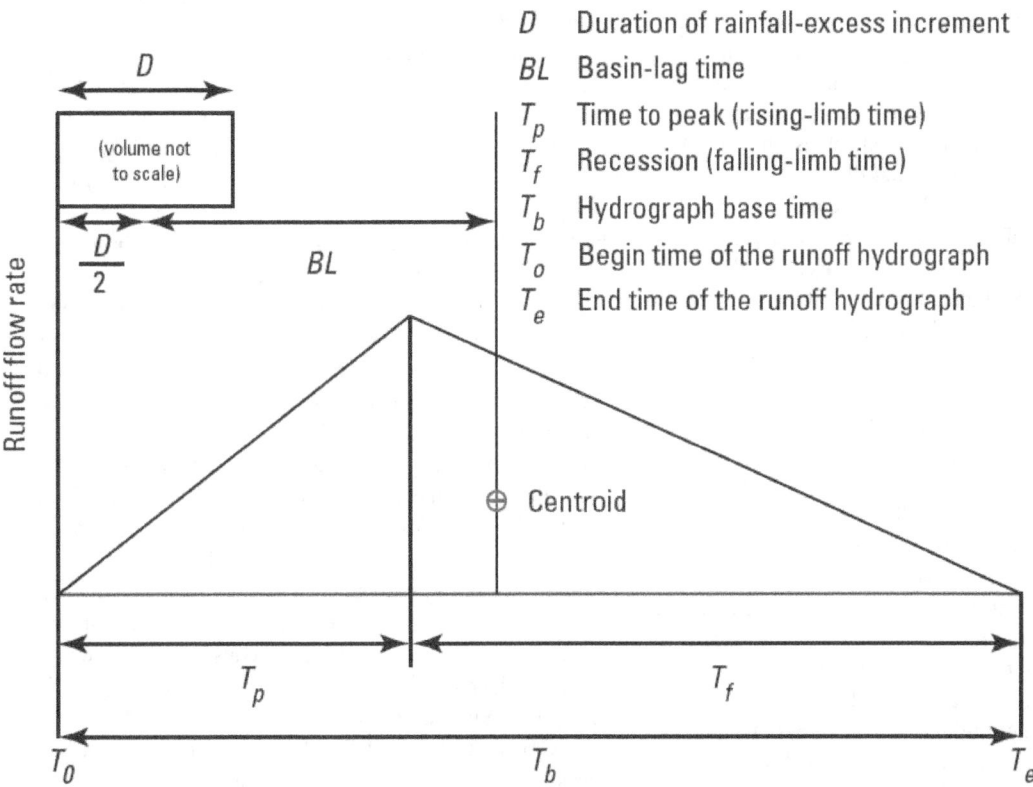

Figure 14. Time factors for a triangular storm-event hydrograph (Modified from Kent, 1973).

where

R_c is the cumulative proportion of the total runoff at time T_i,

T_i is any selected time step within the runoff hydrograph,

T_o is the begin time of the runoff hydrograph,

T_e is the end time of the runoff hydrograph, and

T_p is the peak time of the runoff hydrograph.

If the begin time is set to 0, the end time (T_e) is equal to the duration of the runoff hydrograph T_b (fig. 14). The time to peak is commonly calculated as a function of one-half the duration of rainfall-excess increment ($D/2$) plus a basin lagtime (BL, in hours). In SELDM, the duration of rainfall-excess increment is set equal to the stochastic precipitation duration. Although there are many definitions of the basin lagtime in the literature (Rao and Delleur, 1974; Linsley and others, 1975; Chow and others, 1988; Fang and others, 2005), SELDM defines the basin lagtime as the time from the center-of-mass (centroid) of rainfall excess to the centroid of the corresponding runoff hydrograph. This definition was

selected because the regression equations that are used to calculate the basin lagtime from drainage-basin properties were developed primarily with data from U.S. Geological Survey runoff studies, which are based on the center-of-mass definition (Granato, 2012). The relation between the time to peak T_p (in hours) and the time to the centroid T_c (in hours) of the runoff hydrograph is a function of the ratio (R_f, which is unitless) of the duration of the falling limb to the rising limb of the hydrograph. For a triangular hydrograph this may be calculated as

$$T_p = 3\left(\frac{(D/2)+BL}{R_f+2}\right). \qquad (20)$$

The BL in equation 20 commonly is defined as a characteristic of the basin rather than a characteristic of individual storms, so this variable is fixed in the analysis. SELDM, however, uses Monte Carlo methods to generate random precipitation-event durations (D) and R_f, which are values calculated using input statistics; as a result, T_p is calculated as a random variable, a property which is consistent with observed hydrographs (Granato, 2012). Precipitation-event durations (D) are generated from synoptic precipitation statistics (selected or input on the synoptic storm-event precipitation-statistics form) with a two-parameter exponential distribution. The upstream hydrograph recession-ratio values (R_f) are generated by using user-defined values (input on the upstream-basin form) with a triangular distribution (table 1) (Granato, 2010, 2012). The R_f value for the highway site, however, is fixed at a value of 1 (an isosceles triangle with equal rising- and falling-limb durations); this method is consistent with the rational method for runoff hydrographs for small highly impervious sites (Granato, 2010).

Granato (2012) developed multiple linear regression equations to estimate the values of basin lagtime used in SELDM from drainage-basin properties of the highway site and the upstream basin (which are entered on the highway-site and upstream-basin forms). These drainage-basin properties are defined in detail in appendix 2. The basin lag factor (BLF), which is the basin length in miles divided by the square root of the channel slope in feet per mile was selected as the primary physiographic variable. One multiple linear regression equation that included the BLF and the BDF (appendix 2) for calculating basin lagtimes was developed on the basis of data from 493 sites documented in 22 different studies (the primary dataset). Another multiple linear regression equation for calculating basin lagtimes from the BLF and the TIA was developed on the basis of data from 896 sites documented in 37 different studies (the secondary dataset). Both datasets are comprehensive representations of the characteristics of potential highway sites and upstream basins. Basin drainage areas range from 0.000116 mi^2 (about 0.074 acre) to 1,477 mi^2 in both datasets. The median drainage areas are 4.1 and 3.8 mi^2 in the primary and secondary datasets, respectively. BDF values range from 0 to 12 with a median of 5 in the primary dataset. TIA values in both datasets range from 0 to 100. These regression equations are

$$BL = 0.967 \times BLF^{0.571} \times (13 - BDF)^{0.681} \qquad (21)$$

and

$$BL = 0.499 \times BLF^{0.601} \times (100 - 0.99 \times TIA)^{0.443} . \qquad (22)$$

The structure of these equations and the selected explanatory variables were informed by results of many other studies (Granato, 2012), including the USGS nationwide regression equations (Sauer and others, 1983). Selected values of the regression equations are shown with the source data in figure 15. Equation 21 is used to calculate the basin-lagtime values for the highway site or upstream basin by SELDM if a BDF in the range from 0 to 12 is entered in the highway-site form or the upstream-basin form. To use equation 22 to calculate the basin-lagtime values for the highway site or upstream basin as a function of the BLF and TIA, enter a BDF value of -1 in the appropriate form. A BDF value of -1 has

no physical meaning; entering this value directs the model to use TIA (equation 22) rather than the BDF (equation 21) to calculate BL.

Triangular hydrograph-recession-ratio (R_f) statistics also are needed to calculate the time to peak T_p (equation 20) and the duration of the runoff hydrograph T_b for each individual storm event. The minimum (R_{f-Min}), most probable (R_{f-MPV}), and maximum (R_{f-Max}) values may be estimated visually, by using literature values, or by calculating values from storm hydrographs from nearby hydrologically similar basins. Granato (2010) documented a range of values of R_f from the literature that might be suitable for an analysis. He found that hydrograph-recession studies are not common in the literature because most stormflow studies focus on the basin lag and magnitude of the peak flow to provide information for flood control. However, he found that qualitative estimates of R_f ranged from 1 to 12 in several studies with values that were attributed to basin characteristics such as basin size, slope, and the degree of development. Granato (2012) developed a computer program and several spreadsheets to implement methods for fitting triangular hydrographs and calculating recession-ratio statistics. These methods were developed to estimate R_f statistics using instantaneous streamflow data available from the USGS instantaneous data archive (U.S. Geological Survey, 2012).

Granato (2012) developed triangular hydrograph-recession ratio statistics by using instantaneous streamflow data from 32 USGS streamgages draining portions of Massachusetts and 9 USGS streamgages in other areas of the country (fig. 16). In this dataset, the minimum recession ratios ranged from 1 to 1.77 with a median of 1.02 and a mean of 1.15; thus, the minimum recession ratios are well characterized using a value of 1. The most probable value (MPV) of recession ratios ranged from 1 to 3.52 with a median and mean of 1.85. This median compares well with the median of 1.67 developed from average curvilinear flood hydrographs from USGS studies in different areas of the country. The maximum recession ratios ranged from 2.66 to 11.31 with a median of 4.32 and a mean of 4.76.

Granato (2012) found that correlations between recession ratios and basin characteristics were weak; this result, which was consistent with the findings of other hydrograph recession studies in the literature (Shamir and others, 2005; Shuster and others, 2008), precluded development of meaningful predictive equations. Quantitative R_f selections are not possible through use of the information documented by Granato (2012), but qualitative estimates can be made on the basis of hydrologic similarity. Comparison of basin characteristics at the site of interest with basin characteristics for streamgages in the R_f dataset compiled by Granato (2012) may inform the choice of R_f statistics that are greater than or less than the median values in figure 16. For example, development is weakly associated with decreased MPVs of R_f values, whereas forested areas and onstream impoundments are weakly associated with increased MPVs.

Dilution Factors

SELDM uses the stochastically generated highway stormflows, BMP discharges, and concurrent upstream stormflows to calculate dilution factors for each storm. The dilution factor is the ratio of highway runoff (or BMP discharge) to downstream flow. The dilution factor is calculated as

$$DF = \frac{HQ}{DQ} = \frac{HQ}{(HQ+UQ)},$$ (23)

where

    DF is the dilution factor, which is dimensionless;

    HQ is the highway runoff or BMP-discharge volume, in cubic feet (ft^3);

    DQ is the downstream stormflow concurrent with the highway runoff or BMP discharge, in ft^3; and

    UQ is the upstream stormflow concurrent to the highway runoff or BMP discharge, in ft^3 (Driscoll and others, 1990b).

The dilution factor can vary from 0 to 1 as the highway runoff increases in proportion to the upstream flow. A dilution factor near 0 indicates that highway runoff is a negligible portion of the downstream flow. A dilution factor of 1 indicates that the downstream flow is all highway runoff. The dilution factor increases as dilution decreases; Driscoll and others (1990b) defined the dilution factor in this way to prevent division by 0 errors in the 1990 FHWA runoff model. SELDM calculates dilution factors for highway runoff with and without BMP modification. A BMP that extends highway stormflows will decrease the dilution factor (increasing dilution) by incorporating a larger portion of the total upstream stormflow as concurrent stormflow (fig. 13). The highway and BMP dilution-factor outputs will be equal if there is no BMP flow modification.

The dilution-factor output provides a quick initial assessment of the risks for water-quality excursions with and without BMP treatment. For example, examination of the dilution-factor file for each of several highway-stream crossings can be used to identify the streams with the highest potential for excursions. Similarly, if a highway with many outfalls is parallel to a stream, information about the cumulative upstream drainage and pavement areas at each outfall can be used to run SELDM. The dilution-factor file for each outfall can be used to identify the point along the stream with the highest potential for excursions. In either case, this information can be used to allocate resources for a detailed analysis of flows, loads, and concentrations at the most critical site(s).

Highway and Upstream Stormwater Concentrations and Loads

Estimates of highway and upstream stormwater concentrations and loads are needed for using a mass-balance approach to predict the potential effects of runoff on receiving waters (Warn and Brew, 1980; Di Toro, 1984; Schwartz and Naiman, 1999; Granato, 2010). The mass-balance approach for storm-event analyses is based on the sum of the loads from the highway and the upstream basin. These loads are the product of stochastically generated random runoff concentrations and flows. The concentration of runoff constituents from the highway can be modeled as stochastic random values or stochastic dependent values. The concentration of runoff constituents from the upstream basin can be modeled as stochastic random values, stochastic dependent values, or stochastic transport-curve values. The statistics for generating these constituent concentrations are based on available data.

Sources of Water-Quality Data

Water-quality modeling methods in SELDM are designed to support the FHWA step-by-step decision tree for water-quality-assessments (Sevin, 1987; Cazenas and others, 1996; Federal Highway Administration, 1998). This process starts with an initial assessment on the basis of available data and proceeds to more detailed analyses if the risk of an adverse effect is unacceptable to decisionmakers. SELDM uses regional water-quality statistics to facilitate generation of initial planning-level estimates. If necessary, initial estimates can be refined with water-quality statistics based on available data collected at nearby hydrologically similar sites or at the site of interest. Basing the initial analyses on available data is a prudent approach because collecting sufficient water-quality data to effectively characterize conditions at a given site is expensive, difficult, and time consuming. Furthermore, regional estimates may be more robust for predicting environmental variables at unmonitored sites than measurements from a relatively short-duration site-specific sampling program, unless this program generates enough data to represent conditions at the site of interest, characterizes the full range of discharges, and is not affected by short-term natural or anthropogenic factors (Hughes and Larsen, 1988; Hosking and Wallis, 1997; Vogel and others, 1998; Robertson and others, 2001; Shirazi and others, 2001; Jenerette and others, 2002). The results of short-duration site-specific sampling programs also may not characterize conditions that are likely to occur after a highway project is complete.

Highway-Runoff-Quality Data

Nationally, the Highway Runoff Database (HRDB) is the primary source of highway-runoff statistics and data for use with SELDM (Granato and Cazenas, 2009). The HRDB application is designed as a data warehouse in which to document data and information from highway-runoff

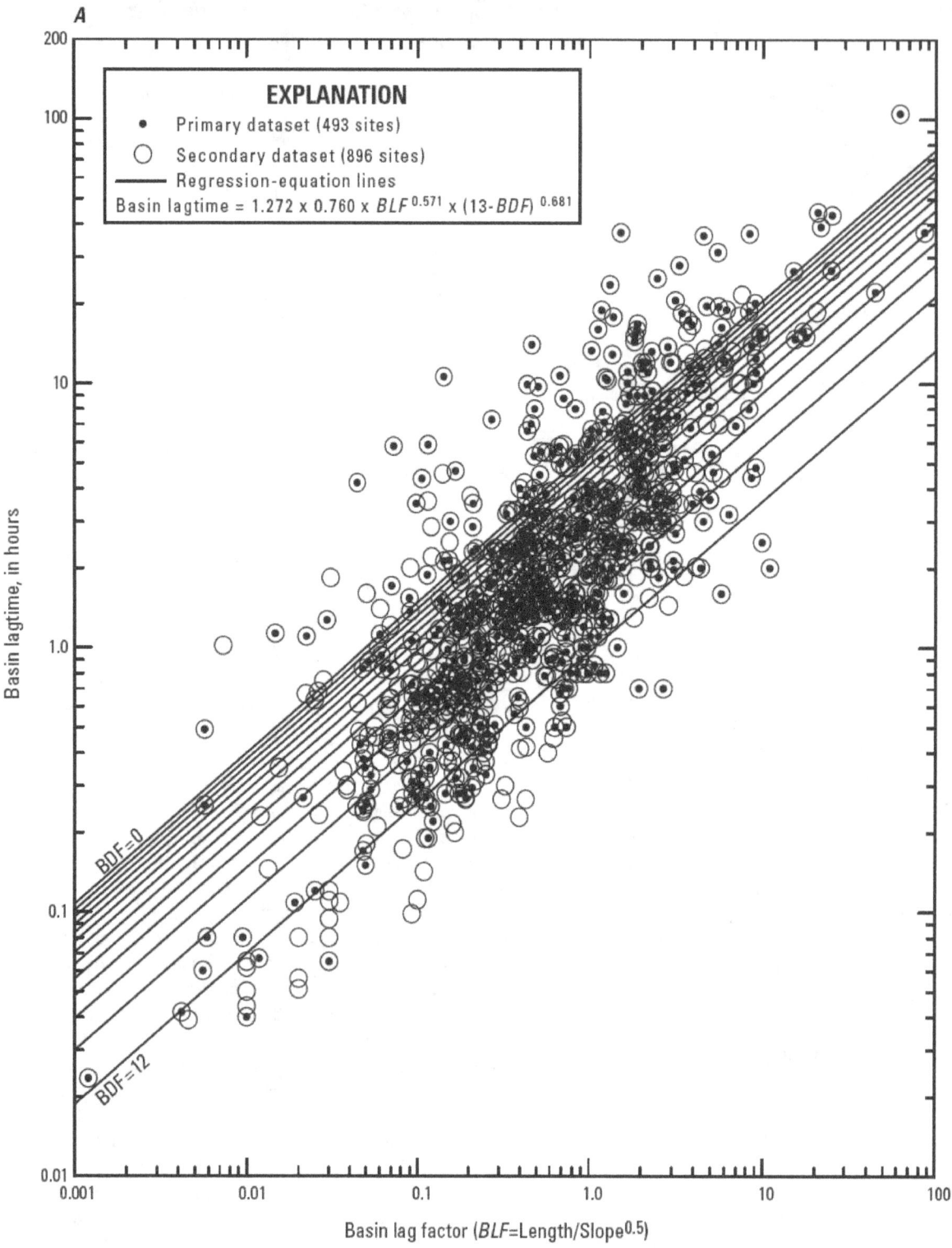

Figure 15. The basin-lagtime data and regression equations based on the basin-lag factor (BLF) and *A*, the basin development factor (BDF) with the primary dataset, or *B*, the total impervious area (TIA) with the secondary dataset.

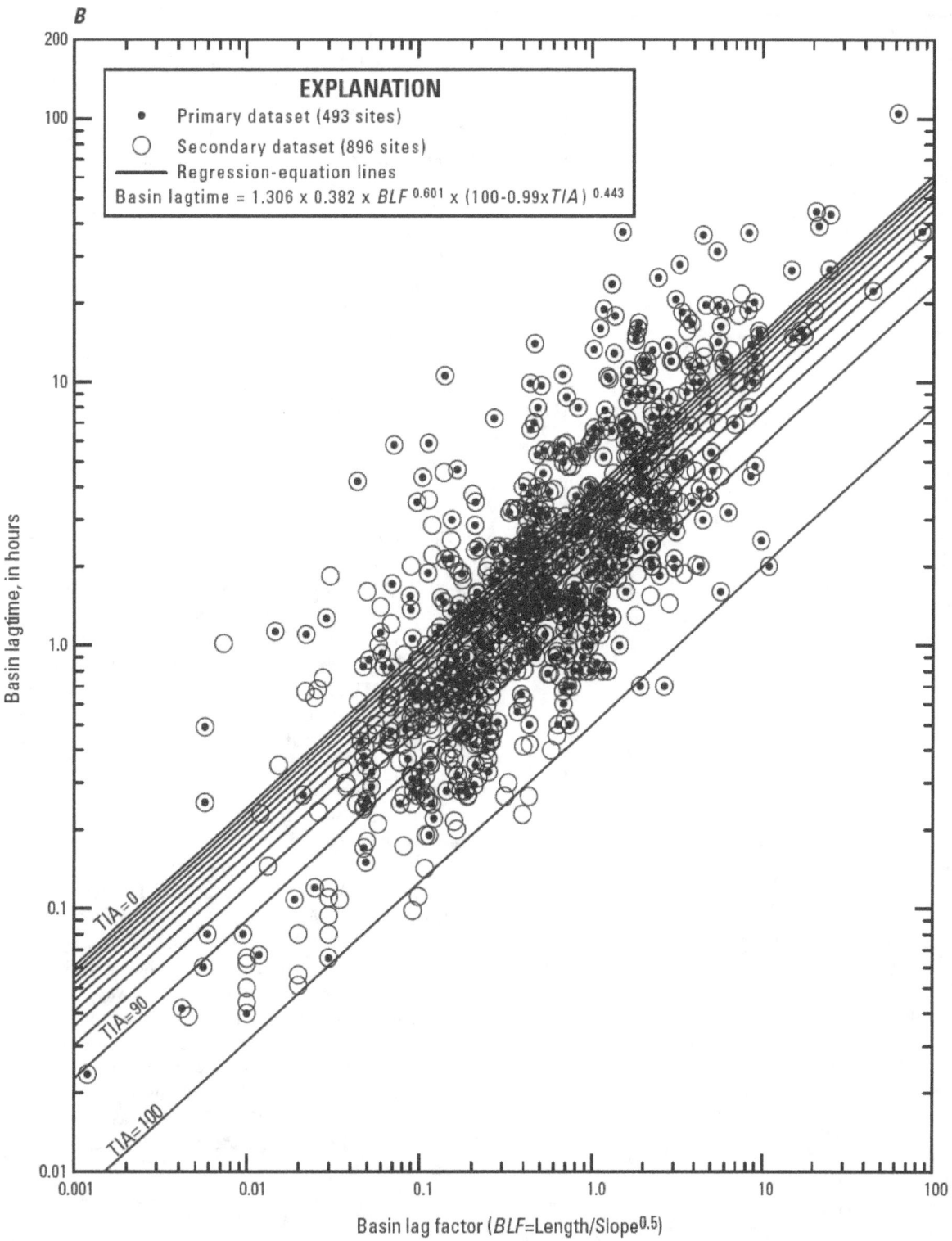

Figure 15.   The basin-lagtime data and regression equations based on the basin-lag factor (BLF) and A, the basin development factor (BDF) with the primary dataset, or B, the total impervious area (TIA) with the secondary dataset.—Continued

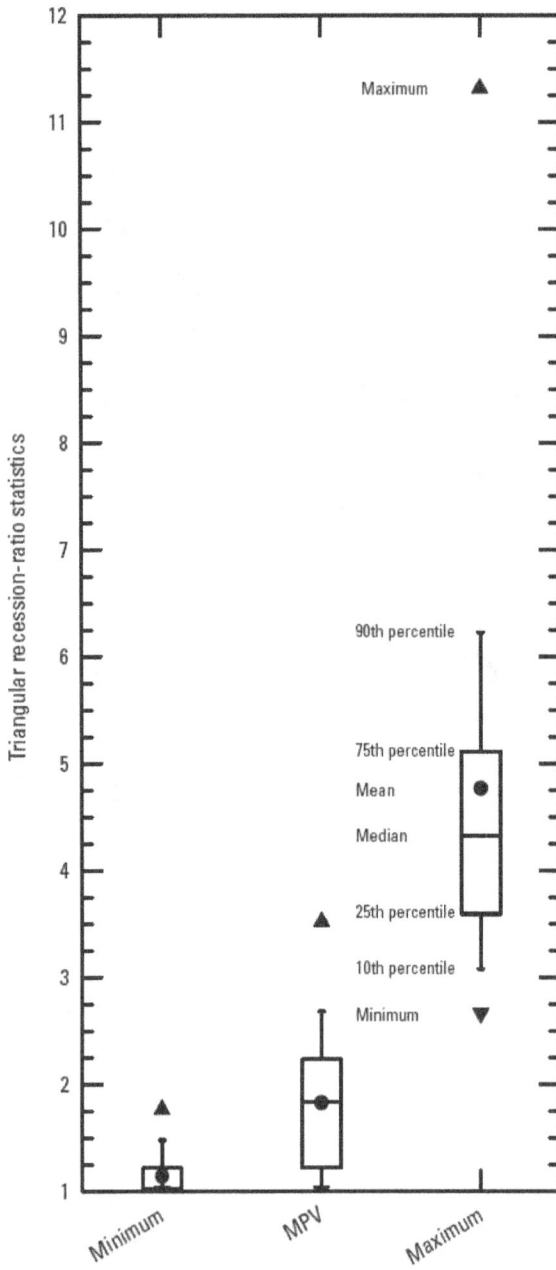

Figure 16. Distributions of the minimum, most probable value (MPV), and maximum for the statistics of the best-fit triangular hydrograph-recession ratio estimated from 20 or more storm-event hydrographs from each of the 41 streamgages in the multistate dataset documented by Granato (2012).

monitoring studies and as a preprocessor for highway-runoff data for use in the SELDM application. Available highway-runoff data provide the basis for defining runoff quality and quantity at monitored sites and predicting runoff quality and quantity at unmonitored sites. Version 1.0.0 of the HRDB included data from 2,650 storms for 39,713 EMC measurements of more than 100 water-quality constituents monitored at 103 sites in the conterminous United States and documented in 7 highway-runoff datasets (Granato and Cazenas, 2009). Smith and Granato (2010) added data from Massachusetts and Ohio to produce version 1.0.0a of the database. This newer version includes 54,384 EMC measurements of 194 water-quality constituents monitored at 117 sites during 4,186 storm events.

The HRDB application also is designed to be a preprocessor for use with SELDM (Granato and Cazenas, 2009). Most common data-manipulation tasks can be accomplished with the GUI of the HRDB or several predefined queries by users with only a cursory knowledge of Microsoft Access®. The database application provides standard and robust estimates of population statistics for highway-runoff data. The HRDB calculates the mean, standard deviation, and skew of the data and the logarithms of data for the random water-quality simulations. It uses accepted methods for calculating these statistics for datasets with one or more values that are below detection limits. The HRDB also is designed to export paired water-quality data in a format that can be used with the KTRLine program (Granato, 2006) to calculate regression statistics defining dependent water-quality relations for highway-runoff modeling.

Additional highway-runoff data also may be available from six state departments of transportation. The HRDB program contains data collected in California through 2004 and in Washington through 2005, but these states have ongoing stormwater-research programs (Gersib, 2011; McGowen, 2011). A study of highway runoff is currently under way in Oregon (Fletcher, 2011). DOTs in North Carolina (Wagner and others, 2011), South Carolina (Conlon and Journey, 2008), and Wisconsin (Horwatich and others, 2011) recently completed comprehensive highway-runoff monitoring studies with the USGS. These data could be integrated into the HRDB to facilitate runoff-quality analyses.

Version 3 of the National Stormwater Quality Database (NSQD) is another potential source of highway-runoff-quality data (Pitt and others, 2008). This database, which is in Microsoft Excel® format, includes data from the National Pollutant Discharge Elimination System (NPDES) from 44 highway sites. The highway-runoff-quality data in this database include 5,454 EMC or grab-sample concentrations of 54 constituents from 734 storm events. This is a large dataset, but Pitt and others (2008) described serious concerns about the reliability and utility of these phase-one stormwater NPDES monitoring data because many different experimental designs, sampling procedures, and analytical techniques were used in the different studies. The NSQD does not include methods to query data or generate statistics beyond the standard Excel features.

Upstream-Water-Quality Data

The national water-quality database compiled by Granato and others (2009) may be the primary source of planning-level upstream-water-quality data that is readily available for use with SELDM. The compilation was based on data available in the USGS National Water Information System Web (NWISWeb), a source of water-quality data that can be used to estimate local, regional, or national water-quality parameters (Mathey, 1998; U.S. Geological Survey, 2002, 2011). As of October 2011, NWISWeb included data for 118,000 stream sites, 2,456 canals, 1,701 ditches, 550 outfalls, 534 wetlands, 211 storm sewers, 145 combined sewers, and 141 pavement sites in the conterminous United States. It also includes stream water-quality data for 2,540 streams in Alaska, 1,031 streams on the Caribbean islands, and 1,480 streams on the Pacific islands. Review of the data, however, indicates that the number of water-quality samples listed for any given constituent is small for many sites. For example, in a review of sediment data in the NWISWeb database, Turcios and others (2010) determined that 30 or more paired suspended sediment and streamflow measurements were available for fewer than 25 percent of the listed monitoring sites. For this reason, regionalization or combining data from nearby sites may be necessary to produce quantitative estimates of water quality at unmonitored sites from available datasets.

Granato and others (2009) selected water-quality-monitoring stations in the conterminous United States at stream sites having a defined drainage area and at least one paired streamflow and water-quality measurement in NWISWeb. A total of 24,581 stations with drainage areas ranging from 0.002 to 1,140 mi^2 were identified and cataloged for retrieval of water-quality data from NWISWeb (fig. 17). The percentages of the selected sites with drainage areas less than 0.1, 1, and 10 mi^2 are 0.7, 5.4, and 28.9, respectively. These basins were not screened for land-use characteristics or impervious fractions because this analysis was designed to provide generalized estimates of ambient receiving-water quality by ecoregion without incorporation of site-specific characteristics. In comparison, the focus of the USEPA analysis effort has been to define reference or minimally impacted conditions by selecting reference sites or choosing the lowest quartile of data from all sites in a region (U.S. Environmental Protection Agency, 2000; Robertson and others, 2001). However, relatively few sites have been monitored to determine reference conditions. For example, Biesecker and Leifeste (1975) characterized water quality at 57 unaltered sites. Similarly, Smith and others (2003) extrapolated estimates of natural background concentrations of nutrients in streams and rivers of the conterminous United States from the water-quality data for 63 stations

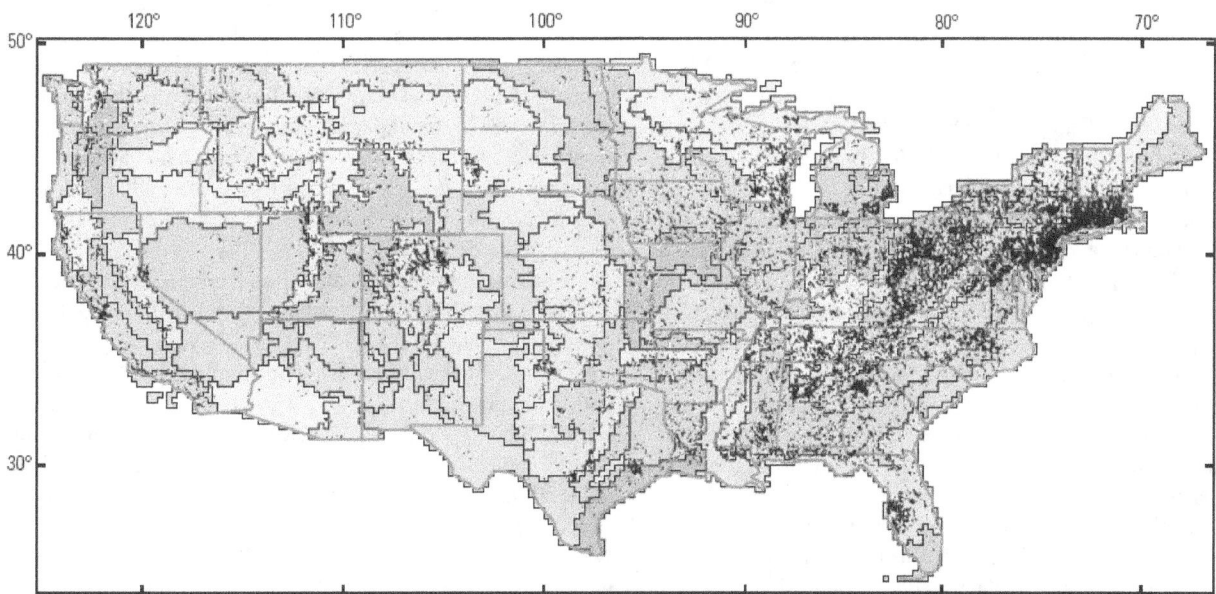

Figure 17. The spatial distribution of 24,581 water-quality-monitoring stations (black dots) with drainage areas less than 1,050 square miles within U.S. Environmental Protection Agency (2003) Level III ecoregions (colored polygons) that have been discretized to a 15-minute grid in the conterminous United States (geographic projection). Ecoregions are identified on the plate useco.pdf on the CD–ROM accompanying this report.

that met all of the USEPA criteria for minimally impacted reference basins. Use of reference basins that represent natural background concentrations, however, may lead to underestimation of water-quality constituent concentrations commonly found in streams receiving highway runoff. Thus, data from reference basins may, on a regional basis, underrepresent the potential for adverse water-quality effects from highway runoff. Conversely, selection of data from sites that are substantially affected by anthropogenic activities may, on a regional basis, over represent the potential for adverse water-quality effects from highway runoff.

Granato and others (2009) developed methods to derive order-of-magnitude planning-level estimates of EMCs in receiving waters at unmonitored sites in the conterminous United States. The methods also may be used to obtain and interpret more quantitative site-specific data. Granato and others (2009) used data-mining and analysis techniques to identify and compile more than 1,876,000 paired streamflow and water-quality measurements that included 21 constituents commonly measured in highway and urban runoff studies and were made between October 1, 1900 and September 30, 2004 (table 4). Granato and others (2009) documented these techniques and the associated computer programs used to obtain and analyze the data (fig. 18). The first step of the data-compilation process is to download paired concentration and streamflow data from NWISWeb with the National Water Information System Wizard (NWiz) developed for this project. Water-quality monitoring stations were identified in NWISWeb and grouped according to ecoregions by using GIS software. Once stations were identified, three programs were used to download and filter the data. These programs are NWiz, the National Water Information System Site Cleaner (NWISSC), and Relational DataBase File Processor (RDBP). The data selected for further analysis was imported into the Surface-Water-Quality Data Miner (SWQDM), which is a relational database that can be used to associate the paired concentration and streamflow data with an ecoregion or any user-specified site location in the conterminous United States. Granato and others (2009) used the SWQDM with the KTRLine program to develop the water-quality transport curves by ecoregion. Transport curves are provided in SELDM for total (unfiltered) phosphorus (parameter code p00665), total hardness (parameter code p00900) and suspended sediment concentrations (parameter code p80154).

Although the dataset compiled by Granato and others (2009) is extensive, a number of limitations apply to use of the dataset. Most of the USGS data probably are not EMC values because most samples were collected for USGS status and trends studies. Methods for collection, processing, and analysis of samples have changed during the period of record. The data include the effects of trends caused by local changes in land use and potentially by a number of regional and national processes through time. The variability in regional water-quality estimates incorporates such at-site variations as well as site-to-site variations within each region. However, the data compilation and interpretation methods described in this report may be used with other information, such as

local land-use data, for more selective regional or local data analysis.

The NSQD database, which contains a substantial amount of NPDES data, also is another potential source of water-quality data that can be used to model upstream stormflow quality (Pitt and others, 2008). The runoff-quality data in this database include 75,291 EMC or grab-sample measurements (including 91 water-quality constituents) from 7,474 storm events monitored at 476 runoff-monitoring sites. These data are comparable to highway-runoff values because they are EMC or grab-sample measurements rather than the instantaneous streamflow and concentration data available in NWISWeb. These data, however, may not be representative of the water quality in many upstream basins because they are limited to relatively small drainage areas with predominantly highly impervious and homogenous land uses. For example, the nonhighway sites in this database have drainage areas ranging from about 0.001 to 16 mi^2, but about 58.5 percent of these sites have drainage areas less than 0.1 mi^2 (65 acres), and 95.25 percent have drainage areas less than 1 mi^2 (640 acres). About 72 percent of the NPDES monitoring sites cataloged by Pitt and others (2008) have TIA values equal to or exceeding 30 percent of the drainage area, a value commonly defined as the threshold for designating urban areas. About 46 percent of these sites have TIA values equal to or exceeding 50 percent of the drainage area, a value commonly defined as the threshold for designating ultraurban areas (Shoemaker and others, 2000). About 57 percent of the drainage areas include one land-use category, about 23 percent two land-use categories, and about 20 percent more than two land-use categories.

Many of the upstream basins that could be analyzed in SELDM may be larger, have lower TIA values, and be more diverse than these NPDES monitoring basins. For example, Falcone and others (2010) cataloged basin properties for 6,785 USGS streamgages, many of which are at road crossings. Fewer than 1 percent of the upstream basin areas to these streamgages are less than 1 mi^2. About 98.6 and 99.9 percent of these areas have TIA values that are less than 30 and 50 percent.

Data also may be available from the USEPA STOrage and RETrieval (STORET) database (U.S. Environmental Protection Agency, 2005b). The STORET Web site includes legacy data collected from the 1960s through 1999 and an option for a modernized STORET system with data collected since 1999. As of 2006, the STORET database did not include data for many states, including parts of Alabama, California, Idaho, Illinois, Indiana, Massachusetts, Mississippi, Nevada, New Mexico, New York, Oregon, Texas, Virginia, and Washington. The STORET database also includes data from wells, wastewater-treatment plants, USEPA Superfund sites, landfills, and mine-discharge points, which are important on a local scale but may skew estimates of ambient receiving-water quality for a regional analysis. The USGS NWISWeb and the USEPA STORET databases use common definitions and formats to provide a common view of data for the two systems. The methods described in this report may be adapted for selective use with STORET data.

Table 4. Paired water-quality and streamflow measurements made between October 1900 and September 2004 compiled by Granato and others (2009) from data for 24,581 water-quality-monitoring stations throughout the conterminous United States.

[Pcode, parameter code; USEPA, United States Environmental Protection Agency]

USEPA PCode	Number of samples	Number of stations	PCode definition
			Physical properties, major ions, solids and sediment, and bacteria
p00340	15,273	959	Chemical oxygen demand, high level, water, unfiltered, milligrams per liter
p00403	129,666	7,753	pH, water, unfiltered, laboratory, standard units
p00900	107,289	7,290	Hardness, water, unfiltered, milligrams per liter as calcium carbonate
p00940	206,341	10,936	Chloride, water, filtered, milligrams per liter
p00530	84,364	2,397	Solids, suspended, water, milligrams per liter
p00535	28,831	525	Solids, volatile suspended, water, milligrams per liter
p80154	275,950	7,477	Suspended sediment concentration, milligrams per liter
p31501	22,998	1,052	Total coliform, M-Endo MF method, immediate, water, colonies per 100 milliliters
			Nutrients
p00600	50,160	2,820	Total nitrogen, water, unfiltered, milligrams per liter
p00630	137,231	4,936	Nitrite plus nitrate, water, unfiltered, milligrams per liter as nitrogen
p00665	246,403	8,169	Phosphorus, water, unfiltered, milligrams per liter
p00671	113,896	5,594	Orthophosphate, water, filtered, milligrams per liter as phosphorus
p00680	83,212	3,513	Organic carbon, water, unfiltered, milligrams per liter
			Metals
p01027	38,352	2,986	Cadmium, water, unfiltered, micrograms per liter
p01113	84	8	Cadmium, water, unfiltered, recoverable, micrograms per liter
p01034	40,855	3,244	Chromium, water, unfiltered, recoverable, micrograms per liter
p01118	34	8	Chromium, water, unfiltered, recoverable, micrograms per liter
p01042	40,563	2,978	Copper, water, unfiltered, recoverable, micrograms per liter
p01119	554	28	Copper, water, unfiltered, recoverable, micrograms per liter
p01045	77,899	5,864	Iron, water, unfiltered, recoverable, micrograms per liter
p01051	35,585	2,771	Lead, water, unfiltered, recoverable, micrograms per liter
p01114	408	19	Lead, water, unfiltered, recoverable, micrograms per liter
p01055	69,248	5,511	Manganese, water, unfiltered, recoverable, micrograms per liter
p01067	18,805	1,891	Nickel, water, unfiltered, recoverable, micrograms per liter
p01074	34	3	Nickel, water, unfiltered, recoverable, micrograms per liter
p01092	52,131	3,837	Zinc, water, unfiltered, recoverable, micrograms per liter
p01094	468	27	Zinc, water, unfiltered, recoverable, micrograms per liter

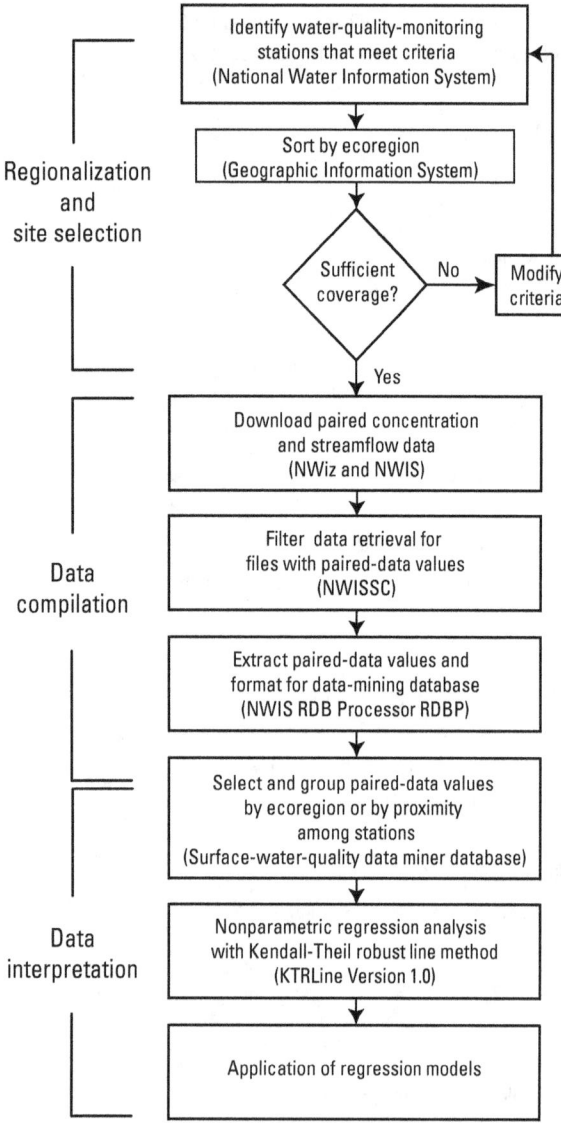

Figure 18. The steps (and associated software) used for regionalization and site selection, data compilation, and data interpretation to define relations between surface-water discharge and concentrations of selected water-quality constituents.

If the sediment-associated fraction dominates the total concentration of a constituent in receiving waters, then the suspended sediment concentration can be used with the concentration of that constituent in sediment to approximate upstream concentrations (Breault and Granato, 2003; Mahler and Van Metre, 2003). The USGS National Water-Quality Assessment Program (NAWQA, accessible at http://water.usgs.gov/nawqa/) has produced many reports characterizing the concentrations of trace elements and organic

chemicals in fine-grained streambed sediments within study areas throughout the United States. The USGS also provides extensive datasets characterizing the trace-element chemistry of soils and stream sediments in the National Geochemical Database (Smith, 2006) and the National Geochemical Survey (Grossman and others, 2008). Rice (1999) compares trace-element concentrations in urban and rural areas across the conterminous United States and provides this dataset on the Internet. These data can be used with SELDM by specifying dependent water-quality relations in the model.

Random Water-Quality Modeling

The SELDM random water-quality option provides the means for stochastically generating a population of constituent concentrations from statistics calculated from the average, standard deviation, and skew of available concentration data. Random water-quality definitions can be defined for highway runoff and upstream stormflow. For the upstream basin, the random water-quality option may be preferred if an analysis shows that relations between constituent concentrations and streamflow (the transport-curve option) are weak. Stormflow water-quality data commonly are characterized and modeled as being from a lognormal distribution, but other distributions also are used (Athayde and others, 1983; Di Toro, 1984; Driscoll, Shelley and others, 1989; Driscoll and others, 1990b; Van Buren and others, 1997; Novotny, 2004; Burton and Pitt, 2002; Maestre and others, 2004; National Research Council, 2008).

SELDM uses the frequency-factor method (equation 1) to generate random constituent concentrations. SELDM can be used to generate concentration data approximating a truncated normal distribution, a truncated Pearson Type III distribution, a lognormal distribution, or a log-Pearson Type III distribution. SELDM also can be used to model concentration data that fit selected exponential or gamma distributions, which are special cases of the Pearson distribution (Bobee and Ashkar, 1991). The water-quality statistics may be entered in the form of arithmetic values or the natural (base e) or common (base 10) logarithms of data. Misspecification of the data type on the input form can cause large errors in generated values. The average and standard deviation of the data or the logarithms of the data are used to generate values with a normal or lognormal distribution, respectively; for these distributions, the coefficient of skew must be set equal to 0. Use of the Pearson Type III or log-Pearson Type III distribution is not an explicit choice offered by the interface; to use these distributions, a nonzero-skew value must be entered with the associated data-transformation option. If arithmetic values are used, SELDM produces data following a truncated normal distribution or a truncated Pearson Type III distribution because the use of untruncated distributions can produce negative concentration values, which have no physical meaning. SELDM truncates these results by fitting such values to a logarithmic lower tail, but the results may not accurately model the input statistics. A detailed description of methods for entering random water-quality statistics is provided in the section of appendix 4 on the random water-quality statistics input form.

Selection of statistics from robust and representative datasets is important for modeling runoff quality. The input statistics that are selected can have a substantial effect on the TMDL allocation assigned to a DOT and the potential number of water-quality excursions that may be modeled. The standard errors of the input statistics are a function of the variability of the data and the number of samples in the dataset used to calculate the statistics (Haan, 1977; Chow and others, 1988; Stedinger and others, 1993; Burton and Pitt, 2002). Although SELDM will typically generate data for 500 to 2,500 storms (depending on precipitation statistics and the random seed), the uncertainty in input statistics is based on the size of the measured dataset, not the size of the modeled dataset. Furthermore, because SELDM generates many more storms than are characterized in the measured dataset by using a theoretical distribution, the range of values in the modeled output is expected to exceed the range of the values in the measured dataset. Thus, the effect of selected statistics may be accentuated in the tails of the data. For example, the effect of the skew coefficient on modeled concentration data for suspended sediment is shown in figure 19. The statistics in the measured dataset were calculated by using the HRDB from 27 highway-runoff EMC values collected along State Route 2 in Littleton, Massachusetts (Smith and Granato, 2010). The percentiles, which were calculated by using the Cunnane plotting-position formula (Helsel and Hirsch, 2002), range from 2.21 to 97.79, whereas the range of plotting positions for the stochastic populations, 0.038 to 99.962, is larger.

Two stochastic populations were generated on the basis of the average and standard deviation of the logarithms of data. The first stochastic population was modeled on a lognormal distribution with a skew of 0; the second was modeled as a log-Pearson Type III distribution with the skew value calculated for the measured dataset. The skew, which is negative, produces a probability plot that is concave downward leading to reductions in the lower-tail and upper-tail values with respect to the lognormal population values. The measured data range from 6 to 406 mg/L, the lognormal values range from 1.83 to 5,020, and the log-Pearson type III values range from 0.447 to 1,000. If the skew of the logarithms of data were positive, the result would be a probability plot that is concave upward leading to potentially large increases in the lower-tail and upper-tail values with respect to the lognormal population values. The number of allowable excursions would depend on the magnitude of highway flows; the effectiveness of the specified BMP; and the magnitudes of upstream concentrations, flows, and loads. All else being equal, however, one would expect higher runoff concentrations to produce a greater number of excursions. A sensitivity analysis using the standard errors of the input statistics will reveal the effect of inputs on water-quality excursions and perhaps the need for a better input dataset. For example, Burton and Pitt (2002) indicate that 25 to 50 EMC

samples may be needed on the basis of the variability of urban and highway-runoff concentrations from numerous datasets.

Specification of statistics for some constituents requires special consideration to ensure that output values are representative of water quality in the environment. For example, the nonlinear pH scale is the negative logarithm of the hydronium ion concentration. pH can range from 0 to 14 (excluding concentrated acid or base solutions), and the pH of most natural river waters is expected to be in the range from 6.6 to 8.5 (Hem, 1992). Similarly, 95 percent of pH values in highway runoff ranged from 5.25 to 8.24 in version 1.0.0a of the HRDB (Smith and Granato, 2010).

Dependent Water-Quality Modeling

The SELDM dependent water-quality option provides the means for stochastically generating a population of constituent concentrations from statistics calculated on the basis of available monitoring data and regression equations characterizing the relations between datasets for two constituents (fig. 20). Dependent water-quality relations can be defined for highway runoff and upstream stormflow. The dependent water-quality option is intended for use if data are limited for a constituent of interest, but the relations between the constituent of interest and other constituents that are well characterized by random statistics or water-quality transport curves are good. SELDM uses the concentration values from the stochastic population of the selected independent variable to generate concentration values for the dependent variable. For highway-runoff constituents, the independent variable must comprise a population of random highway-runoff concentrations. For the upstream constituents, a population of random upstream concentrations or upstream concentrations generated from a water-quality transport curve can be used.

For dependent relations, the input regression-equation statistics are used to calculate a regression result, and Monte Carlo methods for regression equations are used to reproduce the scatter of data above and below the regression line (equation 2). These numbers are stochastically generated on the basis of the assumptions that the residuals are centered on the regression value and are normally or lognormally distributed above and below the regression-line value. If arithmetic regression statistics are used, SELDM produces residual values that follow a truncated normal distribution because use of the lower-tail values can produce negative concentration values, which have no physical meaning. If the calculated concentration is less than or equal to 0, SELDM returns the arbitrarily selected value of 0.002. Thus, output from SELDM that includes multiple concentration values that are equal to 0.002 may indicate that the input statistics produced concentration values less than or equal to 0. Regression equations developed from the logarithms of data are commonly more robust for runoff-quality analysis than regression equations based on untransformed values (Glysson, 1987; Vogel and others, 2005; Granato, 2006; Granato and others, 2009). Misspecification of the data type on the

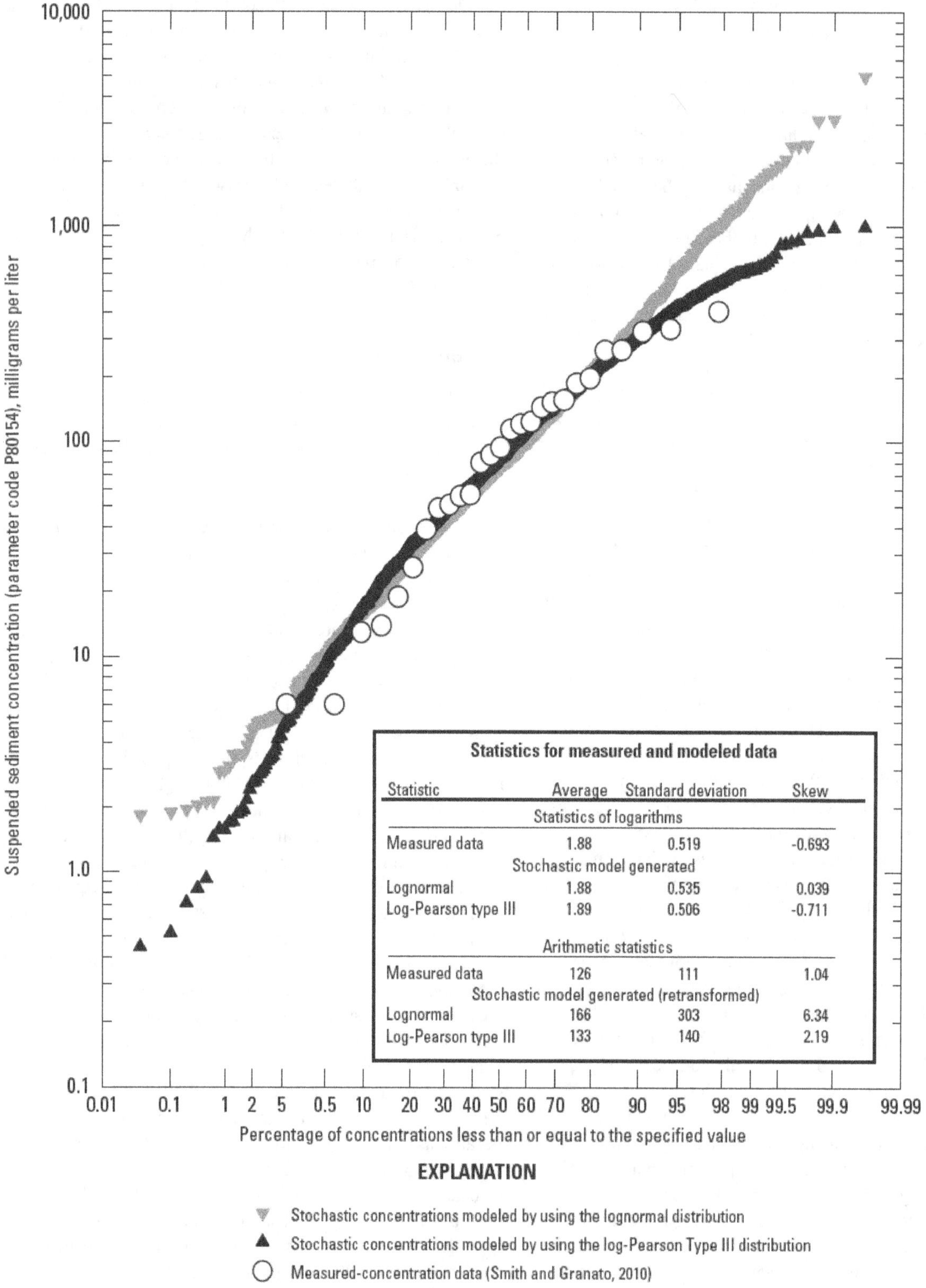

Statistics for measured and modeled data

Statistic	Average	Standard deviation	Skew
Statistics of logarithms			
Measured data	1.88	0.519	-0.693
Stochastic model generated			
Lognormal	1.88	0.535	0.039
Log-Pearson type III	1.89	0.506	-0.711
Arithmetic statistics			
Measured data	126	111	1.04
Stochastic model generated (retransformed)			
Lognormal	166	303	6.34
Log-Pearson type III	133	140	2.19

Figure 19. An example of random water-quality data generated by using statistics from 27 suspended sediment concentrations measured in highway-runoff samples collected along State Route 2 in Littleton, Massachusetts (Smith and Granato, 2010). Stochastic model-generated suspended sediment concentrations were generated by applying lognormal and log-Pearson Type III distributions to statistics from the highway data.

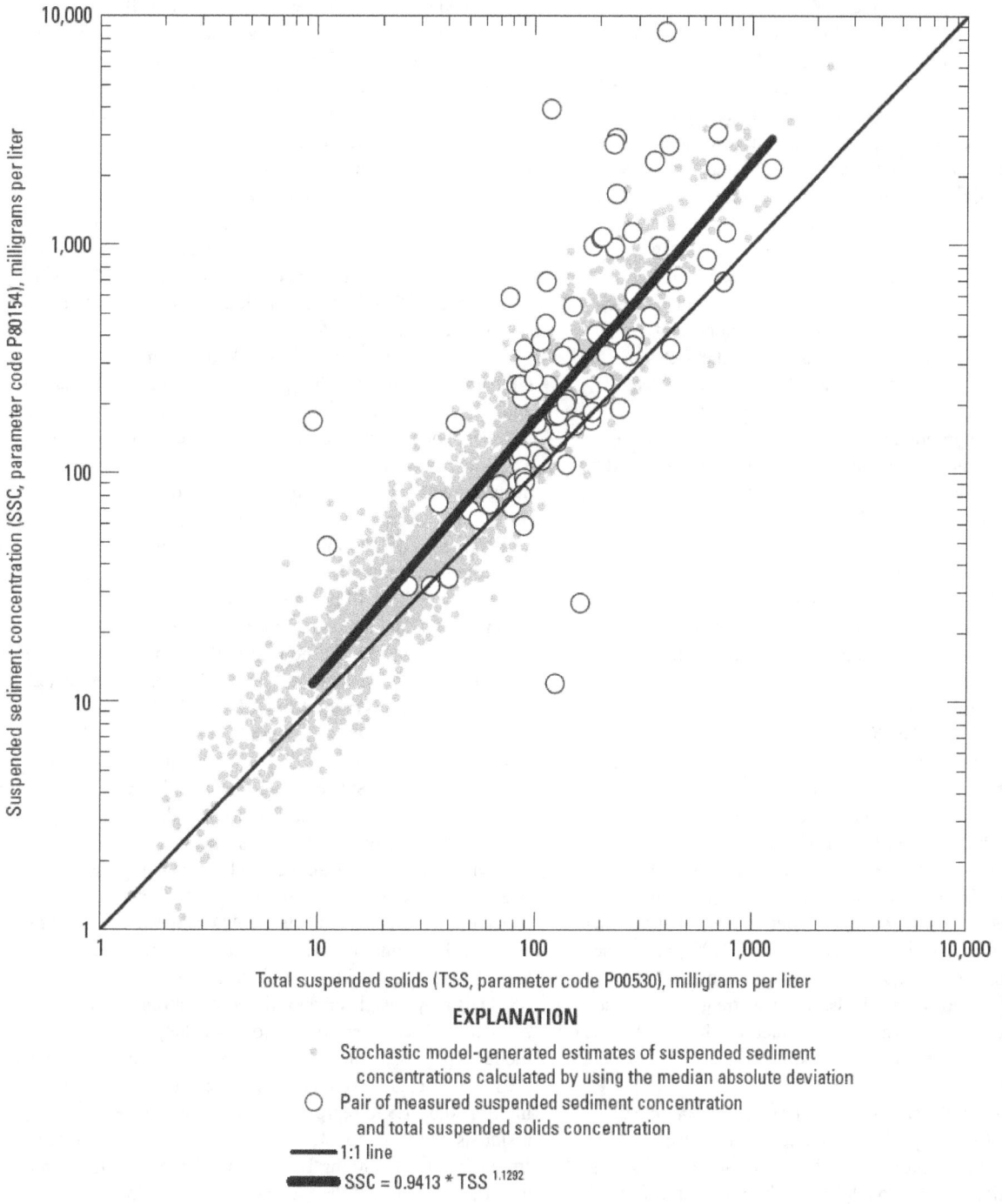

Figure 20.   An example of dependent water-quality data generation using a regression relation for predicting suspended sediment concentrations from total suspended solids concentrations measured in 94 highway-runoff samples (Granato and Cazenas, 2009). Stochastic suspended sediment concentrations are generated by using statistics for stochastic total suspended solids concentrations for nonurban highways (Driscoll and others, 1990a).

input form can cause large errors in generated values. For example, if arithmetic regression statistics are entered, but the logarithmic data type is selected, then data that are orders of magnitude too high may be generated. A detailed description of methods for entering dependent water-quality statistics is provided in the section of appendix 4 on the dependent water-quality statistics and the transport-curve input form. Special constituents like pH require evaluation to ensure that output values are representative of water quality in the environment.

The output-file format of the KTRLine program is designed to facilitate specification of input statistics for SELDM (Granato, 2006). Selection of regression statistics from robust and representative datasets is important for modeling runoff quality because of potential effects on the TMDL allocation assigned to a DOT and the potential number of water-quality excursions that may be modeled. The KTRLine program (Granato, 2006) produces robust regression statistics because it is a nonparametric method that is not heavily influenced by outliers, nonconstant variance of residuals (heteroscedasticity), or assumptions about values of data less than one or more detection limits (Helsel and Hirsch, 2002; Granato, 2006). The KTRLine program calculates the intercept of the line so that the line passes through the medians of both the independent and dependent values. For some datasets, this method may result in residuals that are not centered on the regression line; in this case, the median deviation can be added to the intercept to center the residual values. The KTRLine program also produces a robust nonparametric measure of the variability of residuals to eliminate or reduce the effects of a few far outliers on the residual statistics: the median absolute deviation (MAD). The MAD statistic is about two-thirds of the standard deviation and about one-half of the interquartile range (IQR) for a population of residual values with a normal (or lognormal if transformed values are used) distribution (Helsel and Hirsch, 2002; Granato, 2006; Granato and others, 2009). SELDM, however, uses the MAD value as an approximation of the standard deviation of residuals. If it is warranted, the standard deviation of residuals can be substituted for the MAD value on the dependent water-quality statistics and transport-curve input form.

Granato and Cazenas (2009) provide an example of a dependent water-quality relation to generate suspended sediment concentrations (SSC) from total suspended solids concentrations (TSS) from 94 paired highway-runoff samples. This dependent water-quality relation, which is included as an option in SELDM, is an example of an application of the dependent water-quality method. Granato and Cazenas (2009) derived this relation because most highway-runoff studies collect TSS samples but not SSC samples, and because concerns about the representativeness of TSS concentrations in receiving waters and in highway and urban runoff are substantial. TSS concentrations have been shown to underrepresent the true concentration of sediments in natural waters. Using results from 17,701 paired SSC and TSS samples from across the country, the USGS has

determined that TSS analyses are "fundamentally unreliable for the analysis of natural-water samples" (Gray and others, 2000; U.S. Geological Survey, 2001; Bent and others, 2003). Different methods are used for TSS analyses, but none of the methods provide consistent and repeatable results for natural-water samples. Furthermore, TSS concentrations have been shown to substantially under represent the true concentration of sediments in natural waters. These results also have been substantiated in a number of highway- and urban-runoff studies (for example, U.S. Environmental Protection Agency, 2005a; Guo, 2006, 2007; Landers and others, 2007; Kim and Sansalone, 2008; Ying and Sansalone, 2008; Granato and Cazenas, 2009).

Figure 20 shows the 94 paired-data samples used by Granato and Cazenas (2009), the regression line between TSS and SSC, and a population of 1,571 stochastically generated sediment concentrations. Random water-quality values for TSS were generated by using statistics for data from sites designated as "non-urban" by Driscoll and others (1990a). The measured TSS concentrations range from 9.25 to 1,230 mg/L with a median of 136.5 mg/L, and the associated SSC concentrations range from 12 to 8,580 mg/L with a median of 242.5 mg/L. About 14 percent of the SSC concentrations in the measured data are less than or equal to the associated TSS values. The modeled TSS concentrations range from 0.765 to 2,450 mg/L with a median of 39.5 mg/L, and the associated SSC concentrations range from 0.667 to 11,900 mg/L with a median of 57.7 mg/L. (The graph shown in figure 20 was truncated to values between 1 and 10,000 mg/L). About 18.5 percent of the SSC concentrations in the modeled data are less than or equal to the associated TSS values.

The variability of stochastic suspended sediment concentration estimates above and below the regression line in figure 20 is less than the variability of measured concentrations. The unadjusted MAD value was selected to estimate the standard deviation of residuals because there was some heteroscedasticity in the residuals. Some variation in the residuals may have been an artifact of combining two datasets. Also, the modeled period of record was longer than the modeled period, so modeled values were extrapolated beyond the range of data used to develop the equation. If the MAD is multiplied by 1.5 to represent the full standard deviation of residuals, then the modeled SSC concentrations would range from 0.653 to 16,300 mg/L with a median of 57.2 mg/L, and about 26.5 percent of the SSC concentrations in the modeled data would be less than or equal to the associated TSS values.

The regression line on figure 20 is drawn so that it extends only as far as the range of measured TSS values that were used to develop the regression equation. The use of regression equations to generate data beyond the range of the explanatory-variable data that were used to develop the equations (extrapolation) can produce biased results (Haan, 1977; Helsel and Hirsch, 2002; Granato, 2006). For example, in developing regression equations between specific conductance and concentrations of chloride, Granato and Smith (1999) found that the regression relation was affected by the presence of other major ions in solution if chloride concentrations were less than 130 mg/L. They also found that the regression relation was affected by physical and chemical changes in solution properties with increasing concentrations of solutes if chloride concentrations were above 580 mg/L. For these reasons, evaluation of the applicability of dependent relations and the potential physical and chemical processes that may affect relations beyond the range of available data would be prudent for users of SELDM. The use of data from other sites or other regions may be instructive for such comparisons.

The regression relation in figure 20 appears to be consistent with the theory of sediment transport. The regression equation predicts that TSS concentrations will be greater than SSC concentrations for TSS values below 1.6 mg/L. Theoretically, this is impossible because SSC is a measure of all solids in the sample, but the two measures of suspended solids are expected to converge at low concentrations (Gray and others, 2000); in this case, the differences are well within measurement errors. The regression relation predicts increasing bias with increasing sediment concentrations. This trend is consistent with theory because the higher energy flows that are necessary to transport high sediment loads also tend to mobilize more of the large-grained fraction of sediment that is not well characterized by TSS methods (Gray and others, 2000).

Upstream Water-Quality Transport-Curve Modeling

The SELDM transport-curve option provides the means for stochastically generating a population of constituent concentrations from upstream stormflow values by using regression-equation statistics (fig. 21). Water-quality transport curves are an accepted method for characterizing water-quality variation with streamflow (Biesecker and Leifeste, 1975; O'Connor, 1976; Glysson, 1987; Vogel and others, 2005; Granato, 2006; Landers and others, 2007; Granato and others, 2009). The transport-curve option was developed for use in SELDM because concentrations of many constituents commonly vary as a result of washoff and dilution processes in receiving waters. Concentrations of sediment and sediment-associated constituents (selected nutrients, organic compounds, and trace elements) commonly increase with increasing streamflow above a base-flow threshold because these constituents are mobilized from the stream or land surface during storms (washoff). For example, Landers and others (2007) found that concentrations of sediment, total phosphorus, total metals, and bacteria increased with increasing stormflow. Concentrations of dissolved constituents commonly decrease with increasing streamflow above a base-flow threshold because these constituents, which are commonly associated with groundwater and point-source discharges, are diluted by less concentrated storm runoff. Landers and others (2007) showed that total dissolved solids

decreased with increasing stormflows. Some constituents with multiple sources may show both increasing and decreasing trends with increasing flow. For example, Landers and others (2007) showed that total phosphorus was diluted by increasing base flows but increased with increasing stormflows in one basin with a wastewater-treatment plant. O'Connor (1976) used data to estimate the base-flow threshold for different sites, but Landers and others (2007) showed that the ranges of base flow and stormflow may overlap because some base flows during wet periods of the year may exceed flows from small storms during dry parts of the year.

SELDM uses the stochastic population of upstream stormflows with a water-quality transport curve to generate concentration values for the specified constituent. A water-quality transport curve, however, will provide only one unique value of concentration for each streamflow value rather than the random distribution of concentrations that are characteristic of water-quality data above and below the transport curve. The input regression-equation statistics are used to calculate the regression relation, and Monte Carlo methods for regression equations are used to reproduce the scatter of data above and below the regression line (equation 2). Arithmetic regression statistics can be used, but relations between streamflow and constituent concentrations commonly are modeled with the logarithms of data because these data can vary by orders of magnitude, and logarithmic regression equations commonly improve linearity, reduce heteroscedasticity, and will not produce constituent concentrations that are less than 0 (Glysson, 1987; Vogel and others, 2005; Granato, 2006; Granato and others, 2009). For this reason, regression equations developed from the logarithms of data are commonly more robust for runoff-quality analysis than regression equations developed from arithmetic values. A detailed description of methods for entering dependent water-quality statistics is provided in the section of appendix 4 on the dependent water-quality statistics and transport-curve input forms. Special constituents like pH require evaluation to ensure that output values are representative of water quality in the environment.

Granato and others (2009) used the KTRLine program to develop water-quality transport curves for total phosphorus (parameter code p00665), total hardness (parameter code p00900), and SSC (parameter code p80154) for each of the 84 ecoregions in the conterminous United States to facilitate generation of robust multisegment regression relations. Total phosphorus and suspended sediment concentrations were selected for analysis because these constituents are commonly identified as constituents of concern. SSC were selected rather than TSS concentrations because the USGS has determined that TSS analyses are not appropriate for characterization of sediment concentrations in receiving waters (Gray and others, 2000; U.S. Geological Survey, 2001; Bent and others, 2003). Total hardness was selected because it is a dissolved constituent that is important for calculating water-quality criteria for metals (U.S. Environmental Protection Agency,

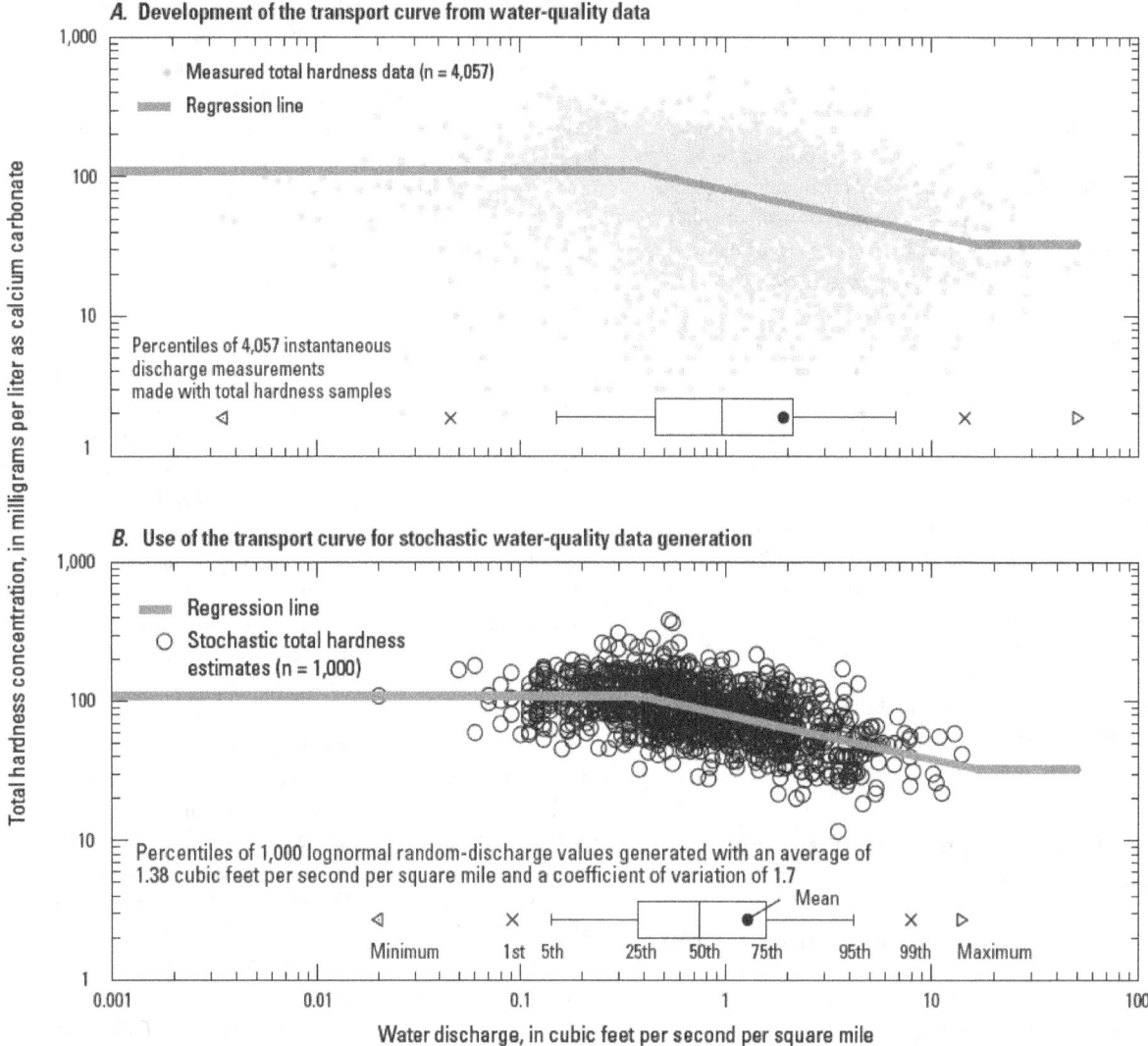

A. Development of the transport curve from water-quality data

B. Use of the transport curve for stochastic water-quality data generation

Figure 21.   Examples of *A,* development, and *B,* use of a three-segment water-quality transport curve for stochastic generation of total hardness concentration data from regional-average daily-flow statistics for ecoregion 67 by use of a random-error component based on the median absolute deviation of water-quality data in each discharge range.

2002a). These three transport curves are provided as a default selection in SELDM.

The interpretive methods used to generate these regional planning-level estimates are based on two assumptions. The first assumption is that data from multiple stations in an ecoregion can be combined to develop a water-quality transport curve that represents ambient water quality at unmonitored sites in that ecoregion. This assumption, however, does not imply that a single regional transport curve is adequate to characterize local variability that may be caused by explanatory factors within the region, but rather that differences among regions of the United States may be characterized by separate transport curves.

The ecoregion-level estimates are intended for an initial screening-level analysis, which may be followed by more detailed analysis with site-specific data if needed. The second assumption is that these transport curves, which were developed on the basis of instantaneous streamflow and concentration data available in NWIS, can be used for order-of-magnitude planning-level estimates of ambient EMCs in receiving waters. Concentrations of dissolved and suspended constituents commonly vary with streamflow in hysteresis loops during a runoff event (O'Connor, 1976; Glysson, 1987; House and Warwick, 1998). Constituents from washoff commonly exhibit higher instantaneous concentrations for a given discharge on the rising part of the hydrograph than on the falling part, whereas constituents from groundwater discharge, which are diluted by increasing runoff, commonly have lower concentrations for a given discharge on the rising part of the hydrograph than on the falling part. A transport curve fitted through instantaneous measurements of streamflow and constituent concentrations collected on both parts of the hydrograph will, by definition, represent the central tendency of measured concentrations (Glysson, 1987; Granato and others, 2009). Therefore, use of the event-mean flow (calculated as the total upstream stormflow volume divided by the duration of the highway runoff or BMP discharge) as the explanatory variable in the transport curve for a given constituent will produce a planning-level estimate of the EMC for that constituent. These assumptions were necessitated by the scale of a national synthesis, the scope of this study, and limitations in available data. The data compilation and interpretation methods described by Granato and others (2009), however, may be used with other information, such as EMC values and local land-use data, for more selective regional or local data analysis.

Granato and others (2009) provide an example of the development and application of a three-segment water-quality transport curve for total hardness on the basis of data from ecoregion 67 (the Ridge and Valley ecoregion used in the I–81 highway-site example provided in SELDM). In this example (fig. 21A), paired measurements of total hardness and streamflow (solid gray dots) were analyzed to determine a three-segment water-quality transport curve (solid black line) with the KTRLine program (Granato, 2006). The total hardness dataset for ecoregion 67 includes 4,057 paired measurements of instantaneous streamflow and total hardness concentrations. The boxplot along the X-axis of figure 21A indicates that samples were collected over more than four orders of magnitude of streamflow. The different segments in the three-segment model also indicate that the statistical properties of a water-quality dataset may depend on the streamflows at which samples were collected (fig. 21A). If samples are not collected throughout the full range of streamflows, regression-line estimates and residual statistics may not be adequate to quantify the range of concentrations that can be found in a region.

Granato and others (2009) used this regional water-quality transport curve with regional flow statistics to generate a population of concentration values for ecoregion 67 (fig. 21B). In this case, 1,000 streamflow values were generated in a lognormal distribution from regional streamflow statistics, which are characterized by the boxplot along the X-axis of figure 21B. Application of the water-quality transport curve (solid black line) and a lognormal distribution of residuals, generated by using the MAD value for each segment, yields a population of total hardness estimates (open circles) above and below the line. Comparison of the regression line, the stochastic data estimates, and the data from ecoregion 67 (figs. 21A and B) indicates that use of this total hardness transport-curve with a stochastic component may be sufficient for generating planning-level estimates of

EMC values that are similar to most of the ambient water-quality measurements.

Granato and others (2009) used total hardness as an example because variations in total hardness with streamflow may affect the application of water-quality criteria for the hardness-dependent trace elements (U.S. Environmental Protection Agency, 2002a). The criteria maximum concentrations or acute criteria (CMCs) for copper may decrease by a factor of 3 (from about 14.7 to 4.67 μg/L) because of the change in hardness between the first and second inflection points on this three-line transport curve. However, the potential dilution over this range of flows increases by a factor of 45 between these streamflow values. Random variation in total hardness concentrations may be an important factor in ecoregion 67 because total hardness and associated CMC values for copper can vary substantially at any given streamflow value. For example, the randomly generated values of total hardness vary from almost 400 to about 30 mg/L for streamflows below $1.0 \text{ ft}^3/\text{s/mi}^2$; the corresponding CMC values for copper could range from about 50 to 4 μg/L at any given streamflow in this range.

Runoff Modification by Best Management Practices (BMPs)

SELDM uses a simple stochastic statistical model of BMP performance to develop planning-level estimates of runoff-event characteristics rather than a complex theoretical or physical model. This statistical approach can be used to represent a single BMP or an assemblage of BMPs. The SELDM BMP-treatment module has provisions for stochastic modeling of three stormwater treatments: volume reduction, hydrograph extension, and concentration reduction. Volume reduction is modeled to represent how BMPs can affect flows and loads from the highway site. Hydrograph extension is modeled to represent how BMPs can increase dilution in receiving waters by extending the duration of runoff from the highway site. Concentration reduction is modeled to represent changes in constituent concentrations that may result from different treatment options.

Using the BMP runoff-control options alters the highway, upstream, and downstream outputs from the model. If BMP volume-reduction statistics are specified, the highway-runoff flows and loads will be affected accordingly. If BMP volume reductions are specified but concentration reductions are not, then the highway-runoff and BMP discharge concentrations will be the same, but the BMP discharge loads and the concurrent downstream loads and concentrations will all be different. If BMP hydrograph extension is specified, the concurrent upstream and downstream flows and loads will be different than for the untreated runoff because the discharge period will be extended to include more of the upstream flow and loads. If BMP concentration-reduction statistics are specified, BMP discharge concentrations and loads will be affected as well as downstream concentrations and loads.

Application of results from BMP monitoring studies is highly uncertain; few studies provide reliable predictions of treatment performance even with large datasets and complex models (Strecker, Quigley, and others, 2001; Wong and others, 2006; Park and others, 2011). Uncertainties arise because of the many categories of BMPs, wide variations in design and construction of BMPs within each category, and wide variations in the operation and maintenance of BMPs once they are installed. Similar BMPs are used at sites with widely varying site characteristics including different precipitation, site hydraulics, constituent characteristics and loads, and total stormwater loads. For example, local soil characteristics can influence the amount of runoff generated by a given storm, the concentrations of sediment in runoff, and the settling rate of the sediments within a BMP. Variations in BMP design also can affect actual and modeled BMP effectiveness. For example, BMP structures may have overflow or bypass structures that have a substantial effect on performance once the BMP volume has been filled. These design features may affect performance only during large storms or storms that occur in rapid succession (Strecker, Quigley, and others, 2001). Uncertainties in effectiveness also arise because BMP monitoring is a complex endeavor that requires a high degree of expertise. Although BMP monitoring protocols have become more standardized, many BMP studies still are conducted individually with different protocols and data-reporting standards rather than as part of a large consistent and coordinated monitoring effort (Jane Clary, Civil Engineer, International BMP Database Project, written commun., May 2011).

Uncertainties in results also are compounded by available sample sizes. Driscoll and others (1979) recommend the collection of 20 to 40 EMC samples to characterize runoff on the basis of the variability of commonly measured runoff constituents. Similarly, Burton and Pitt (2002) indicate that, at a minimum, 25 to 50 EMC samples may be needed. The California Department of Transportation (2009) provides similar examples in their BMP monitoring handbook. These examples indicate that 50 to 113 paired samples may be needed just to detect differences in mean concentrations. In comparison, Leisenring and others (2011) looked at TSS data for 10 types of BMPs in a recent summary of solids-removal data in studies in the International BMP Database. Although TSS is one of the most widely monitored constituents in BMP studies, the average number of paired samples per category ranged from 6 to 16 with a median of about 12 per study. Schneider and McCuen (2006) calculated that monitoring data from about 90 storms would be necessary to fully quantify the hydraulic performance of a stormwater-detention cistern in Maryland on the basis of local precipitation-event characteristics. In comparison, Poresky and others (2011) looked at volume-reduction data from the International BMP Database; they found that the number of storm events ranged from 5 to 173 with a median of about 11 per study. As with other hydrologic data, uncertainty in data related to BMP performance increases when data from one site are extrapolated to estimate conditions at a different site. In addition, small sample sizes limit the ability to select and parameterize statistical distributions for modeling BMPs with data.

In SELDM, volume reduction, hydrograph extension, and water-quality-treatment variables are modeled by using the triangular/trapezoidal distribution and the rank correlation with the associated highway-runoff variables. This family of distributions was selected for modeling BMP performance measures because it can be parameterized by using expert judgment or by fitting the distribution to data if good data are available (Johnson, 1997; Back and others, 2000; U.S. Environmental Protection Agency, 2001; Scherer, 2003; Kacker and Lawrence, 2007). The triangular distribution, which is a special case of the trapezoidal distribution, commonly is suggested when uncertainties in input data that may be used to define a parametric distribution are large (U.S. Environmental Protection Agency, 2001). The triangular/trapezoidal distribution is bounded by a selected minimum and maximum value. When data are uncertain or are limited in scope, use of a bounded distribution reduces the chance that unrealistic output values will be generated by extrapolating a distribution beyond the range of available data.

Volume Reduction

Volume reduction is the practice of retaining, detaining, or routing runoff flows to increase the amount of infiltration, evapotranspiration, or diversion between the pavement and the outfall (Goforth and others, 1983; Schueler, 1987; Urbonas and Roesner, 1993; Wanielista and Yousef, 1993; Young and others, 1996; Adams and Papa, 2000; Burton and Pitt, 2002; National Cooperative Highway Research Program, 2006; Poresky and others, 2011). Volume reduction commonly is a design criterion for BMPs to reduce flood flows, instream erosion, and runoff loads. Features such as flow lengths (for swales) or design volumes commonly are used with moisture-retention estimates and data on local infiltration rates to estimate the volume-reduction capacity of BMPs. Expected storm-event characteristics also are considered in BMP designs because the volume, intensity, and duration of events and the time between storms affect the capacity of the BMP to reduce runoff volumes.

SELDM uses a simple stochastic representation of the net volume reduction from a BMP or series of BMPs. SELDM models flow-volume reductions for generic BMPs as stochastic ratios of inflow to outflow volumes by using the triangular/trapezoidal distribution and the rank correlation with the highway stormflow volume. Volume-reduction statistics for the triangular/trapezoidal distribution can be estimated by using expert judgment or by fitting the distribution to data. Rank correlation with the highway stormflow volume is included in the input statistics because the ratio of outflow to inflow volumes would be expected to increase with increasing storm volumes. For large storms, a smaller fraction of the total inflow may be lost to infiltration or evapotranspiration than for small storms. The different storm characteristics that affect flow reduction, however, also depend on the reduction methods and the design characteristics of the BMP. Because there are many uncertainties in the analysis and application of available BMP data, expert judgment based on knowledge of the hydraulic properties of a BMP design may be the best available information for parameterizing the distribution of volume-reduction ratios.

Many studies have been done to measure and model volume reduction, but accurate categorical determination is hampered by the stochastic nature of antecedent conditions, precipitation, and runoff (Goforth and Heany, 1983; Adams and others, 1986; Driscoll and others, 1986; Schueler, 1987; Driscoll, Shelley, and others, 1989; Urbonas and Roesner, 1993; Wanielista and Yousef, 1993; Young and others, 1996; Adams and Papa, 2000; Huber and others, 2006; Poresky and others, 2011). For example, Emerson and Traver (2008) attributed seasonal two-fold variations in infiltration rates during a four-year period at BMPs in Maryland to changes in the viscosity of ponded water with changes in temperature. Most statistical and deterministic modeling studies are designed to estimate flow reductions for only one type of BMP at a time because of differences among the reduction mechanisms in different BMP designs. Also, most modeling efforts do not characterize the effects of treatment trains (which comprise two or more BMPs in series) because of these complexities.

Despite decades of BMP-monitoring efforts, data are limited for some BMP categories, and substantial uncertainties in the volume-reduction performance of many BMPs remain. In an analysis of flow data in the International BMP Database, Poresky and others (2011) noted that because many older studies were designed to monitor reductions in concentration instead of volume, measurements of volume were made only during the collection of flow-weighted water samples. Thus, flow-duration and volume data may include only the period used for water-quality sampling rather than the complete duration of inflows and outflows. They also noted that the inflow and outflow data from some BMP studies could not be truly paired because these studies measured BMP inputs at only one of many inlets to the BMP or at a reference site that was not associated with the monitored BMP outlet. Poresky and others (2011) emphasize that data are not available for many types of BMPs, and that the level of uncertainty of the available data is high.

The July 2011 version of the International BMP Database (http://www.bmpdatabase.org/) was queried to examine the potential for developing volume-reduction statistics by using available BMP monitoring data. Stormflow data are available for 328 BMP sites; storm-event data indicating the start and end times of inflows and outflows are available for 91 of these sites. The range in the number of storms is 2 to 384, and the median number of storms recorded at each site is 11. A negative flow-extension duration indicates that the period of inflow exceeded the period of outflows. The minimum, average, and maximum flow-extension durations are negative

for about 60, 12, and 3 percent of these 91 sites, respectively. If monitoring data do not include the complete runoff period for inflows or outflows, then the volume-reduction statistics will be biased by the truncated flow volumes. These anomalous results may be caused by the sampling artifacts described by Poresky and others (2011), or they may be the result of data-entry errors. These anomalous results also may reflect the fact that BMP performance estimates also are affected by substantial difficulties in accurately monitoring runoff flows (Strecker, Mayo, and others, 2001; Church and others, 2003; Poresky and others, 2011). Thus, the feasibility and defensibility of transferring BMP monitoring results from one site to a different site is hampered by many uncertainties.

Selected stormflow data from the International BMP Database were used as an example to demonstrate the process for estimating the statistical parameters of the trapezoidal distribution for use with SELDM. First, the inflow and outflow volumes were retrieved from the BMP Database, and the ratios of the outflow to inflow volumes were calculated. Figure 22A is a scatterplot of the inflow and outflow volumes for the San Rafael biofilter, which is a grass strip next to a highway. Inflow volumes range from 68 to 3,424 ft^3, with a median of 424 ft^3. Outflow volumes range from 6 to 1579 ft^3, with a median of 211 ft^3. The ratios of outflow to inflow volumes range from 0.022 to 1.68, with a median of 0.38 (fig. 22B). Then the inflow volumes and the ratios were sorted and ranked to calculate the rank correlation (Spearman's rho) between inflow volumes and the associated ratios (Haan, 1977; Helsel and Hirsch, 2002). The ranks show a weak downward trend with considerable scatter; Spearmans's rho for this dataset is about -0.23 (fig. 22C). Thus, volume-reduction ratios decrease with increasing runoff volumes. This pattern makes sense physically because reductions in flow volumes by BMPs are commonly limited by hydrologic conditions; most BMPs are not capable of removing a large fraction of an extremely high inflow volume. This pattern also makes sense mathematically because larger denominators would tend to reduce ratio values. Although the correlation between the inflow volume and the ratio of outflow to inflow volumes is, by definition, a spurious correlation (Haan, 1977), use of the rank correlation for stochastic data generation reduces the chance that unreasonably large volume reductions may be modeled.

The volume-reduction ratios were then fit to the trapezoidal distribution. The ranks of the ratios were used to calculate the plotting-position percentile for each ratio; for this example, the Cunnane plotting-position formula (Helsel and Hirsch, 2002) was used. The cumulative distribution-function equations for the trapezoidal distribution (Kacker and Lawrence, 2007; appendix 1) were used to calculate theoretical trapezoidal values for each percentile on the basis of trial statistics (fig. 23A). The absolute difference between the data value and the theoretical trapezoidal value was calculated for each percentile. Optimization methods were used to minimize the sum of the absolute differences between the data and the trial statistics. In this example, the Microsoft Excel Solver® in the Analysis Toolpak® was used to optimize

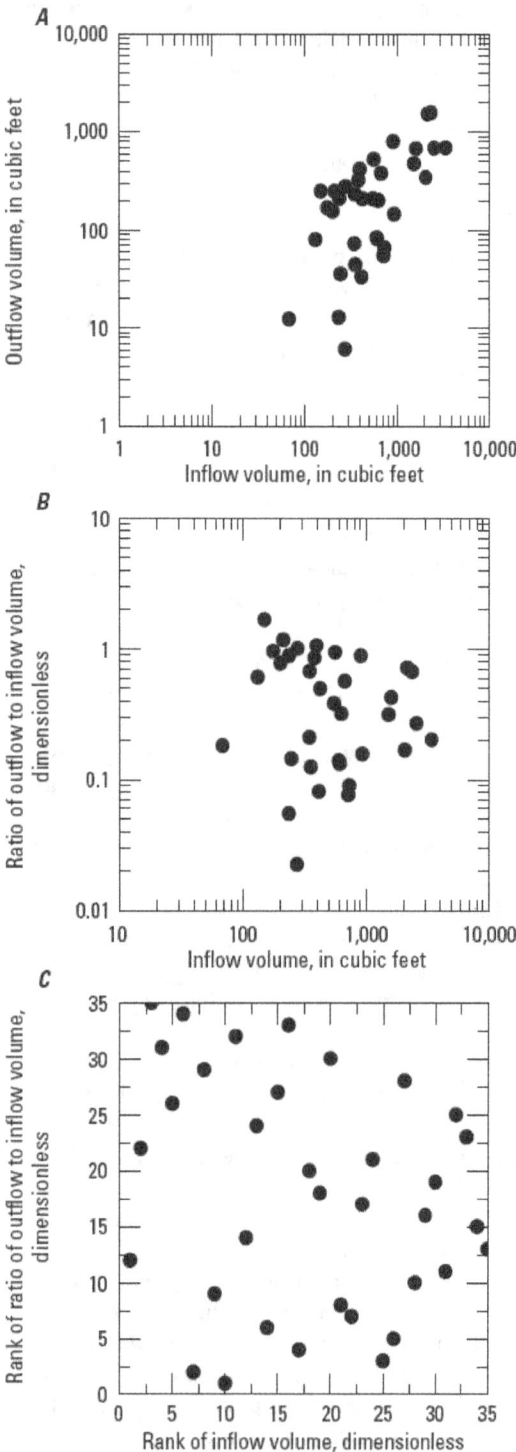

the fit. The regression equations provided by Johnson (1997) also could be used to generate a triangular approximation to the data. Figure 23A shows the data, the statistics for the best fit trapezoidal distribution, and the CDF for the best fit trapezoidal distribution.

The inflow-volume statistics, the trapezoidal-distribution statistics, and the rank correlation values were used with a random-number generator to produce two populations of inflow and outflow values (fig. 23B). The stochastically generated flow values compare well to the to the measured inflow and outflow volumes. Although the populations of outflow volumes look slightly different from the sample of measured outflow volumes, the median and mean of the stochastic values are within the 95-percent confidence limits of the respective sample values. The median value of the measured outflow data is about 202 ft³, the 95-percent confidence interval for the median ranges from 84 to 319 ft³ (calculated by methods described by McGill and others, 1978), the mean value is 328 ft³, and the 95-percent confidence interval for the mean ranges from 187 to 469 ft³ (calculated by methods described by Haan, 1977). In comparison, the median values of the two stochastic populations are about 164 and 161 ft³, and the mean values are about 356 and 348 ft³, both are well within the 95-percent confidence limits of the statistics for the measured outflow volumes. The stochastic inflow and outflow populations have greater ranges than the data because they represent estimates for about 1,500 storms, whereas the monitoring data represent samples from 31 storms. Although the use of the average percent removal has been shown to create problems for predicting BMP performance (Strecker, Quigley, and others, 2001), it is clear, in this example and other test applications, that the use of stochastic ratios with rank correlation to inflows can predict realistic outflows for a site of interest under the assumption that the design features from the example would be properly scaled for the hydrologic characteristics of a site of interest.

The stochastic approach used in SELDM is warranted because there are large uncertainties in available information, and the level of effort required to develop detailed simulation models may be beyond the scope of an initial planning-level estimate. If, however, the initial analysis done with SELDM indicates the potential need for mitigation, then detailed simulation models such as those described by Huber and others (2006) or detailed statistical models such as those described by Adams and Papa (2000) may be used to develop the performance statistics used by SELDM. Furthermore, if the initial analysis without BMP treatment indicates the potential need for mitigation, then SELDM can easily be used to develop the BMP-performance statistics needed to reduce storm loads or the frequencies of water-quality excursions in receiving waters to an acceptable level. This analysis can be done by varying BMP flow-reduction statistics to meet water-quality objectives. Such an analysis may indicate that it is impossible to meet water-quality objectives by using the volume-reduction capabilities of feasible BMP designs.

Figure 22.   Scatterplots showing A, Inflow and outflow volumes from the San Rafael biofilter, B, inflow volumes and the ratios of outflow-to-inflow volumes, and C, the ranks of the inflow volumes and the ratios of outflow-to-inflow volumes. Monitoring data are from the July 2011 version of the International Best Management Practices Database.

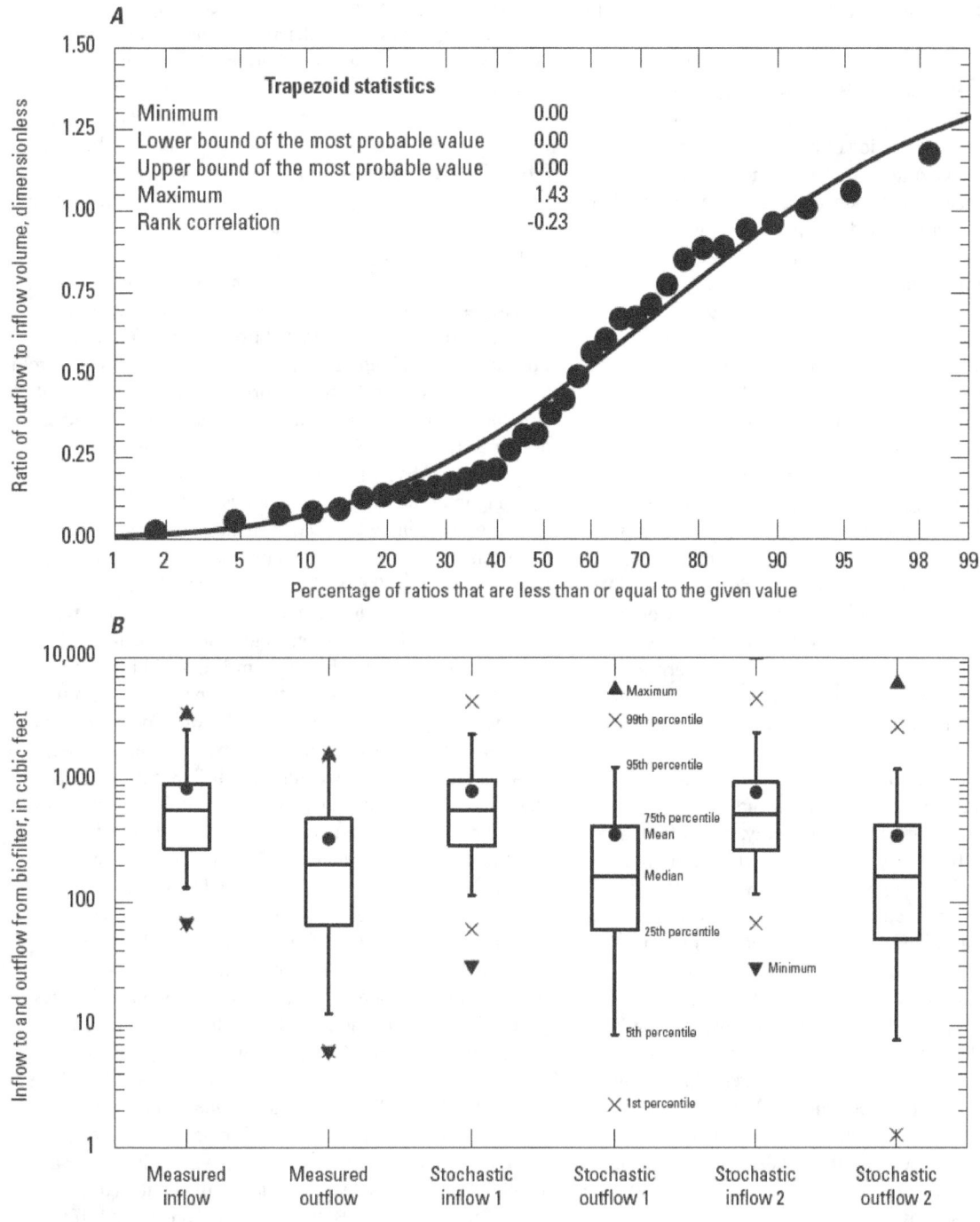

Figure 23. Statistics for estimating stochastic flow reductions for the San Rafael biofilter. The probability plot *A*, demonstrates how the trapezoidal statistics are estimated by fitting the data to the distribution. The boxplots in *B*, showing the measured data and two independent stochastic datasets demonstrate how the stochastic populations of outflow values generated by using the parameterized trapezoidal distribution compare to the original data. Monitoring data from the July 2011 version of the International Best Management Practices Database.

Hydrograph Extension and Concurrent Upstream Flows

Hydrograph extension by BMPs is the practice of slowing the discharge of runoff flows and releasing these flows to the stream over an extended period of time. Hydrograph extension commonly is a design criterion for BMPs to reduce flood flows, to reduce instream erosion and, more recently, to mimic predevelopment stormflow hydrographs. Historically, attempts to optimize detention were done to maximize sediment settling time while minimizing the chance of untreated overflows from subsequent storms (Goforth and others, 1983; Driscoll and others, 1986; Schueler, 1987; Wanielista and Yousef, 1993; Adams and Papa, 2000; Chen and Adams, 2005, 2007; National Cooperative Highway Research Program, 2006). These efforts commonly resulted in extension of the outflow hydrograph. Hydrograph extension also has the added benefit of increasing dilution of runoff from small highly impervious sites. As illustrated in figure 13, extending the duration of the highway-runoff hydrograph can make a big difference in the amount of dilution in a receiving stream, especially in the rising limb of the upstream storm-event hydrograph. SELDM should not be used as a hydraulic design tool, but the output from the program may indicate how the BMP design will affect potential dilution. This information is summarized in the dilution-factor output file from each SELDM analysis, which lists the fraction of highway runoff in the downstream flow with and without BMP treatment.

Hydrograph extension is defined as the duration in hours of discharge from the BMP that occurs after runoff from the highway site has ceased. In theory, runoff from a highway site or a BMP may continue to trickle forth for an extended period of time. In practice, however, the duration of runoff should be defined so that it is truncated at some measurable and meaningful value. For example, minimum precipitation-monitoring depths commonly are about 0.01 in. per hour, which would yield about 0.01 ft^3/s/acre (Church and others, 2003). This threshold, however, may not be measurable at small sites. For example, Smith and Granato (2010) used a storm-monitoring threshold of about 0.009 ft^3/s to distinguish the presence of flow because it was the minimum value that was reliably discernible for a level sensor to detect the presence of flow in 8-in. pipes draining 12,000 to 24,000 square feet (ft^2) of pavement (about 0.03 and 0.016 ft^3/s/acre, respectively).

SELDM models hydrograph-extension times (in hours) as a stochastic variable by using the triangular/trapezoidal distribution and the rank correlation with the highway stormflow volume. Hydrograph-extension times can be estimated by using expert judgment or by fitting the distribution to data. Knowledge of the hydraulic properties of a BMP design may provide the information needed to parameterize the distribution of hydrograph detention times. Alternatively, data from BMP studies in hydrologically similar areas may be used to fit a distribution to available data by using the methods described for volume reduction.

Hydrograph extension is an explicit design feature for some BMPs. For example, detention ponds commonly are designed with outfall weirs and orifices to control the rate of outflow (Goforth and others, 1983; Driscoll and others, 1986; Schueler, 1987; Wanielista and Yousef, 1993; Young and others, 1996; Adams and Papa, 2000; Park and others, 2011). Design standards for BMPs commonly specify drain times of 24–72 hours for a design storm that fills the detention structure. For BMPs with drainage-control structures, therefore, the hydrograph-extension duration would be a function of the volume of water in the BMP at the end of the highway-runoff time. For this type of BMP, the maximum flow extension may be estimated from the design drainage time, and the minimum flow extension may be estimated from the flow formula for the weir or orifice in the drainage-control structure and the minimum highway-runoff volume calculated during a preliminary SELDM modeling run. The upper and lower most probable flow-extension values may be estimated by using total runoff volumes from user-selected flow statistics from the preliminary SELDM modeling run. For example, the mean and median flow volumes may be used as the most probable values of the BMP flow extension. If larger volumes increase the outflow duration, then the correlation between these statistics would be strong and positive. Decreasing the rank correlation coefficient from 1 to 0 will increase the random variation in extension times with respect to storm volume. Using negative correlations between runoff volume and flow extension would invert the expected relation between these variables for a flow-controlled BMP because the flow-extension time would tend to decrease with increasing inflows.

Hydrograph extension is an implicit design feature for some BMPs. If the baseline condition for comparison is a curb-and-gutter highway with a drop inlets and a storm-sewer system draining directly to the stream, then design measures that increase the flow length, lower the drainage slope, increase flow-path roughness, increase flow-path storage, or provide flow resistance at some point in the flow path will extend the outfall hydrograph. For example, a simple application of Manning's equation may indicate the effect of converting a sewer design to a grassy swale (Linsley and others, 1975; Chow and others, 1988). Alternatively, the basin lagtime graph in figure 15 indicates that the basin lagtime for a small highway site (with a basin lag factor of about 0.01) would increase by a factor of about 5 (from about 4 to 20 minutes) as the BDF is reduced from 12 to 3. If the duration of the rising limb of the unit hydrograph equals the duration of falling limb for such sites, bounding estimates for the flow-extension parameters may be 0 to 0.5 hours. Although this is a small increase in duration, it may result in a substantial amount of additional dilution. For example, the average of median daily flows among the 2,783 streamgages in the 2010 SELDM dataset (Granato, 2010) is 0.67 ft^3/s/mi^2. This flow rate, which is expected to be lower than a large percentage of instantaneous stormflows, would result in an additional 1,206 ft^3 (about 34,150 liters) of stormflow per square mile of upstream drainage area. Check dams added to such a swale may further increase the extension time in a predictable manner (Young and others, 1996). Flow extension would be weakly correlated with flow volume for such modifications. Sand filters also may extend the hydrograph in predictable ways: the falling limb of the hydrograph can be estimated for such systems by using Darcy's equation (Linsley and others, 1975; Chow and others, 1988; Young and others, 1996).

The triangular/trapezoidal hydrograph-extension parameters may be fit by using good data if good data are available for a given BMP design. Methods used for estimating flow-extension statistics are similar to methods described for estimating flow-reduction statistics. The July 2011 version of the International BMP Database (http://www.bmpdatabase.org/) was queried to examine the potential for developing hydrograph-extension statistics for BMPs. Stormflow data is available for 328 BMP sites; data indicating the start and end times of inflows and outflows are available for 91 of these sites. Stormflow-duration data are available for 2 to 384 storms per site; the median number of storms per site is 11. However, flow extension estimates made using data from the International BMP Database may under represent actual BMP performance because, as Poresky and others (2011) noted, measurements of volume were made only during the collection of flow-weighted water samples in many older studies. Thus, flow-duration data may include only the period used for water-quality sampling rather than the complete duration of inflow and outflow hydrographs. These concerns are supported by analysis of duration data for the 91 sites: the minimum, average, and maximum flow-extension durations are negative for about 60, 31, and 3 percent of the sites, respectively. Therefore, careful screening of duration data is advisable for developing hydrograph-extension statistics from these data.

Water-Quality Treatment

Water-quality treatment is the use of physical and chemical processes in an attempt to reduce the concentration of runoff constituents in stormflow. Hundreds of BMP studies have focused on water-quality treatment during the past 40 years. Settling and filtration commonly are the primary water-quality-treatment mechanisms that form the basis for reductions in influent concentrations for many constituents in commonly used BMP designs (National Cooperative Highway Research Program, 2006; Clary and others, 2010, 2011; Leisenring and others, 2010, 2011). Increasingly, however, chemical and biological processes are being incorporated into BMP designs to enhance treatment of runoff constituents. Historically, process modeling (for example, methods described by Huber, 2006; and Park and others, 2011), theoretical statistical modeling (for example, Adams and Papa, 2000), and data analysis (for example, Strecker, Quigley, and others, 2001; Barrett, 2005, 2008; and Leisenring and others, 2010, 2011) have been used to examine BMP performance. The water-quality-treatment module in SELDM is designed to

be used with the results from the statistical analysis of BMP data. Alternatively, however, process modeling or theoretical statistical modeling can be used to estimate the statistical parameters used to model BMP performance with SELDM.

Interpretation of available data is complex. Although data are plentiful, uncertainties abound. Differences in site characteristics, BMP designs, precipitation characteristics, influent characteristics, and many other factors affect how BMPs perform from storm to storm and over long periods. Differences in monitoring design, execution, and documentation also introduce uncertainties in available BMP-performance results. For example, Poresky and others (2011) note that volume control can be estimated from BMP-design details, but water-quality-treatment efficiencies commonly show little or no dependence on design details within the random variation of available data.

Two primary measures are commonly used for characterizing and comparing water-quality treatment for different BMPs: the average efficiency and the minimum irreducible concentration. There are various ways for calculating each of these parameters (Strecker, Quigley, and others, 2001; Barrett 2005, 2008; Wright Water Engineers and Geosyntec Consultants, 2007; and Park and others, 2011). The average efficiency is commonly calculated by dividing the average influent concentration by the average effluent concentration or by averaging the individual concentration ratios for sample pairs. Efficiencies for groups of BMPs have been calculated by averaging the average values for individual BMPs or by pooling data and then calculating averages. In some cases, loads are used to calculate flow-weighted averages. Use of average efficiencies to compare BMPs, however, can lead to substantial errors in performance estimates over the range of concentrations expected for a single site (Strecker, Quigley, and others, 2001; Barrett 2005, 2008; Wright Water Engineers and Geosyntec Consultants, 2007; Park and others, 2011). Furthermore, use of average efficiencies can compound errors in performance estimates when the averages are applied to sites with different influent characteristics.

The minimum irreducible concentration is commonly defined as the lowest concentration achievable for a well-designed example of each type of BMP (Schueler, 1996; Barrett, 2005, 2008; Geosyntec Consultants and Wright Water Engineers, 2009). Use of a minimum irreducible concentration reflects the fact that most BMPs will not produce distilled water, so there will be some lower limit that can be achieved with normal BMP unit operations. Although the concept of the minimum irreducible concentration is sound, determining such a value from data may be difficult especially if data are limited, the selected BMP is not characteristic of design standards, or a substantial proportion of the effluent concentrations are below historical detection limits. Schueler (1996) examined available data and settling times to estimate minimum irreducible concentrations. Barrett (2005) developed regression relations between influent and effluent concentrations and interpreted the intercept as a good estimate of the minimum irreducible concentration. The average efficiency and minimum irreducible concentration concepts are linked because an individual BMP receiving runoff that is close to the minimum irreducible concentration may exhibit low efficiencies because the ratios of inputs to outputs are small.

The effluent probability method (EPM) suggested by Strecker, Quigley, and others (2001) is a good diagnostic tool for evaluating BMPs; however, the EPM was not designed to be a predictive tool for estimating effluent concentrations. Furthermore, with a few exceptions, the results of an EPM analysis are not directly transferable to sites with substantially different influent concentrations (Park and others, 2011). For example, an EPM may show a statistically significant performance benefit at sites with high influent concentrations, but no discernible concentration differences at sites with low influent concentrations (Strecker, Quigley, and others, 2001; Barrett, 2005; Park and others, 2011).

Barrett (2005) used regression analysis to determine relations between influent and effluent concentrations and provided 90-percent confidence intervals for effluent-concentration estimates to build predictive models for BMPs and compare performances. He also used the intercept of the equation to estimate the minimum irreducible concentration. This method is useful as an analysis tool, but with parametric regression techniques, one or more erroneous measurements can define a regression relation (Helsel and Hirsch, 2002; Granato, 2006). Furthermore, the regression approach is based on the assumption that data are paired. Strecker, Quigley, and others (2001) indicate that outflow for a particular BMP may have little or no relationship to the inflow for that same event.

Barrett (2008) used the ratio of average influent concentrations to average effluent concentrations as a predictive tool to measure performance and a means of comparing different types of BMPs. His approach differed from the average efficiency approach because he presented results as a series of ellipses that bounded expected performance for different types of BMPs at sites with varying influent concentrations. He demonstrated that BMP performance was a function of influent concentration, so that site characteristics must be considered in the evaluation process.

SELDM uses a simple stochastic representation of the net concentration reduction from a BMP or series of BMPs. SELDM models concentration reductions for a generic BMP as a stochastic ratio of inflow to outflow concentrations by using the triangular/trapezoidal family of distributions, the rank correlation of concentration-reduction ratios with influent highway stormflow concentrations, and a minimum irreducible concentration. Given a population of influent concentrations selected on the basis of highway characteristics, the stochastic ratio provides the variation in outflow ratios and concentrations that are seen in monitoring data. Rank correlation provides the capability to model the relations between influent and effluent concentrations. For example, Barrett (2005, 2008) notes that the effluent concentrations of TSS from sand filters are fairly constant and independent of

influent concentrations. In this case, a strong negative rank correlation between influent concentrations and the ratio of effluent to influent concentrations would tend to assign high ratios to low influent concentrations and low ratios to high influent concentrations and thus reduce the magnitude and variability of effluent with respect to influent concentrations. Applying the minimum irreducible concentration would provide a lower bound for calculated effluent concentrations.

The triangular/trapezoidal concentration-reduction parameters may be estimated by using expert judgment or by fitting the distribution to good data if they are available for a given BMP design. Unlike a regression approach, use of a bounded distribution reduces the chance that unrealistic output values will be generated by extrapolating a relation beyond available data. Although a series of equal lower-bound concentrations set to a minimum irreducible concentration would not be lognormal, such data (above the detection limits) are apparent in many of the probability plots in the statistical summaries from the international BMP database for nutrients (Leisenring and others, 2010) and solids (Leisenring and others, 2011).

Total phosphorus concentration data from the International BMP Database were used to demonstrate the process for determining the statistical parameters of the trapezoidal distribution for use with SELDM. First, the inflow and outflow concentrations were retrieved from the BMP Database, and the ratios of the outflow-to-inflow concentrations were calculated. Figure 24A is a scatterplot of the inflow and outflow concentrations from the Lake Ridge detention pond. Inflow concentrations range from 0.04 to 1.3 mg/L, with a median of about 0.31 mg/L. Outflow concentrations range from 0.04 to 0.43 mg/L, with a median of about 0.21 mg/L. The ratios of outflow to inflow concentrations range from 0.11 to 1.75, with a median ratio of 0.7 (fig. 24B). Then the inflow values and the ratios were sorted and ranked for determination of the percentiles of the ratios and the rank correlation between inflow concentrations and the associated ratios. The Cunnane plotting-position formula (Helsel and Hirsch, 2002) was used to calculate the plotting-position percentile for each ratio for this example. Spearmans's rho (Haan, 1977; Helsel and Hirsch, 2002) was used to calculate the rank correlation between inflow concentrations and the associated ratios. The scatterplot of the inflow concentrations shows that the ratios decrease with increasing inflow concentration; Spearmans's rho for this dataset is about -0.76 (fig. 24C). This makes sense physicochemically because many BMPs are better able to substantially reduce very high than very low concentrations and because it is difficult to improve the runoff quality substantially if inflow concentrations are low (the "clean water in = clean water out" phenomenon described by Leisenring and others (2010, 2011)). This result also makes sense mathematically because larger inflow concentrations (the denominators) would tend to reduce ratio values. Although the correlation between the inflow concentrations and the ratio of outflow to inflow concentrations is, by definition, a spurious correlation (Haan, 1977), use of the rank correlation

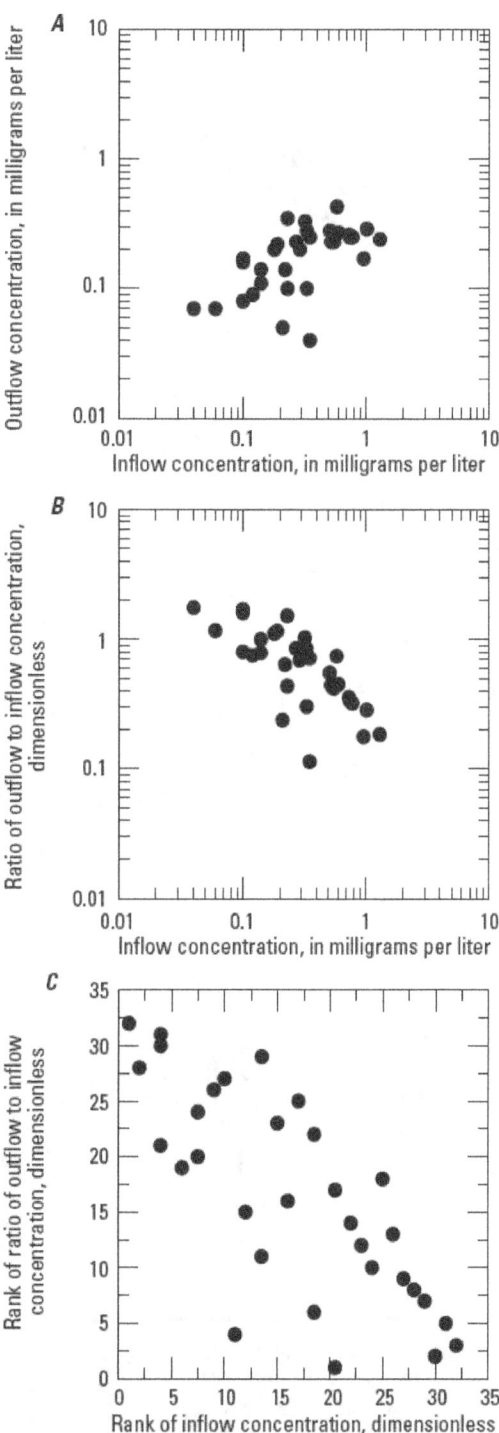

Figure 24. Scatterplots showing *A*, Inflow and outflow concentrations of total phosphorus from the Lake Ridge detention pond in milligrams per liter, *B*, inflow concentrations and the ratio of outflow-to-inflow concentrations, and *C*, the ranks of the inflow concentrations and the ratios of outflow-to-inflow concentrations. Monitoring data are from the July 2011 version of the International Best Management Practices Database.

for stochastic data generation reduces the chance that unreasonably large concentration reductions may be modeled.

The CDF equations for the trapezoidal distribution (Kacker and Lawrence, 2007; appendix 1) were used to calculate values for each percentile on the basis of trial statistics. The absolute difference between the data value and the theoretical trapezoidal value was calculated for each percentile. Optimization methods (in this case, the Microsoft Excel® Solver) were used to minimize the sum of the absolute differences between the data and the trial statistics. The regression equations provided by Johnson (1997) also could be used to generate a triangular approximation to the data. Figure 25A shows the data, the statistics for the best fit trapezoidal distribution, and the CDF for this trapezoidal distribution.

The inflow water-quality statistics, the trapezoidal-distribution statistics, and the rank correlation values were used with SELDM to generate two populations of paired inflow and outflow values. The stochastically generated total phosphorus concentrations compare well to the measured inflow and outflow concentrations (fig. 25B). Although the populations of outflow values look slightly different from the sample of total phosphorus concentration data, the median and mean phosphorus concentrations for the stochastic values are within the 95-percent confidence limits of the respective measured values. The median concentration value (0.21 mg/L) is slightly higher than the mean value (0.20 mg/L) in the measured outflows. The 95-percent confidence interval for the median value of the measured data, however, ranges from 0.16 to 0.26 mg/L (calculated by using methods described by McGill and others, 1978), and the 95-percent confidence interval for the mean value ranges from 0.01 to 0.38 mg/L (calculated by using methods described by Haan, 1977). In comparison, the median values of the two stochastic populations are about 0.17 mg/L, and the mean values are about 0.20 mg/L; both are well within the 95-percent confidence limits of the statistics for the measured outflow concentrations. The stochastic populations have greater ranges than the data because they represent estimates for about 1,600 storms (30 years of record), whereas the monitoring data represent samples from only 32 storms.

The approach used in SELDM is warranted because of the uncertainties in available information and because the level of effort required to develop deterministic simulation models may be beyond the scope of an initial planning-level estimate. If, however, the initial analysis done with SELDM indicates the potential need for mitigation, then detailed deterministic simulation models such as those described by Huber and others (2006) or detailed statistical models such as those described by Adams and Papa (2000) may be used to develop the performance statistics used by SELDM. Furthermore, if the initial analysis indicates the potential need for mitigation, then SELDM could easily be used to develop the BMP-performance statistics for concentration reduction needed to reduce storm loads or the frequencies of water-quality excursions to acceptable levels.

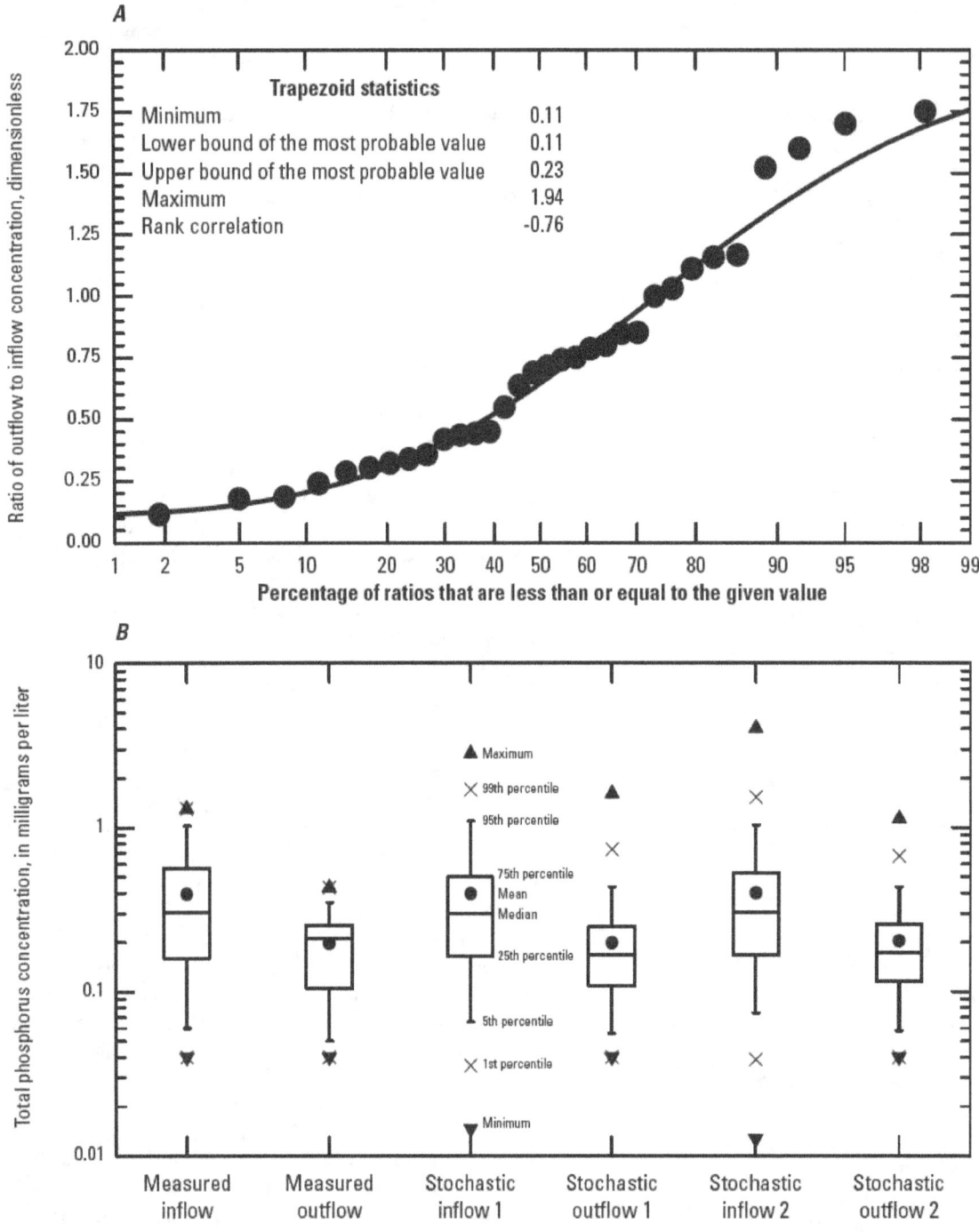

Figure 25.   Statistics for estimating stochastic reductions in total phosphorus by the Lake Ridge detention pond. The probability plot *A*, demonstrates how the trapezoidal statistics are estimated by fitting the data to the distribution. The boxplots *B*, demonstrate how the stochastic populations of outflow values generated by using the parameterized trapezoidal distribution compare to the original data. Monitoring data are from the July 2011 version of the International Best Monitoring Practices Database.

Downstream Stormwater Concentrations and Loads

Downstream water quality is defined by the mixing of highway-runoff loads (concentration times flow) with upstream loads (fig. 1). SELDM generates a random population of flows, concentrations, and loads for the highway, with and without BMP treatment, and for the upstream basin. Highway and upstream loads are added and then divided by the combined flow to calculate downstream concentrations. The upstream loads and flows used for the calculations are the values that are either concurrent with discharge from the highway site or with discharge from the selected BMP. If BMP-treatment options are not defined or selected for a given water-quality pair, then the downstream water-quality values are calculated with the upstream flows and loads that occur during the period of highway runoff. If, however, BMP-treatment options are defined and selected, then the downstream water-quality values are calculated with the upstream flows and loads that occur during the period of BMP discharge.

Water-Quality Pairs

SELDM uses water-quality pairs to specify how the downstream concentrations are calculated. To define a pair, the water-quality constituent must be defined and selected for both the highway runoff and upstream flow. The same highway and upstream constituent selections may be defined as multiple pairs with different short names. To do a sensitivity analysis within one run of SELDM, multiple pairs could be defined by using the same highway and upstream water-quality constituent selections. For example, the same highway and upstream constituent selections could be used to calculate downstream concentrations, flows, and loads as one downstream constituent-pair with BMP treatment and a second downstream constituent-pair without BMP treatment. Similarly, the same highway and upstream selections can be used to calculate different downstream concentrations of concern by defining different constituent pairs with different adverse-effect concentrations. Similarly, the sensitivity of annual average lake concentrations to estimated apparent annual average attenuation factors can be tested by defining different constituent pairs with the same highway and upstream constituent selections but different attenuation factors. SELDM outputs are accompanied by a description of the characteristics of the pair. This description may be used to document the assumptions and sources of data used to define the selections.

Concentration of Concern

SELDM deliberately defines the term "concentration of concern" in a vague manner because different methods may be used for identifying a concentration of concern, and SELDM may be used to address different kinds of water-quality problems. The concentration of concern, however, is designed to represent that portion of the measured or modeled constituent concentration that may have an adverse effect on natural or human water use in the receiving water downstream of the highway outfall. SELDM uses adverse-effect ratios to calculate the concentration of concern as a fraction of the calculated downstream EMC for each storm event. Thus, the concentration of concern is not a single regulatory threshold defining water-quality excursions that would require mitigation measures for runoff control, but instead the component of each modeled EMC that would apply to a single regulatory threshold.

For example, if dissolved nutrients in the water column are of concern, then the adverse-effect ratios may be selected to represent the dissolved (filtered) portion of the modeled total whole-water concentration. If, however, only one component of the dissolved nutrients is of concern, then the adverse-effect ratio would represent the fraction expected to be contributed by only that component. Similarly, if trace metals in the water column are of concern then, depending on the regulatory approach, the adverse-effect ratios may be selected to represent the entire dissolved (filtered) portion of the total whole-water concentration, the filtered portion minus the colloidal portion as a fraction of the total whole-water concentration, or just the free ionic portion of the total whole-water concentration (Tipping, 1994; U.S. Environmental Protection Agency, 1996c; Tipping and others, 1998; Di Toro and others, 2001; Santore and others, 2001). If, however, the dissolved concentrations are being modeled directly, then the adverse-effect ratios would be calculated to represent the fraction of the dissolved concentration that accounts for the free ionic portion. Similarly, if local deposition of metals or nutrients to bed sediments is of concern, and whole-water concentrations are being modeled, then the adverse-effect ratios may be selected to represent the proportion of metals or nutrients expected to be associated with coarse sediments (Breault and Granato, 2003; Buckler and Granato, 2003; Lopes and Dionne, 2003).

An alternate approach to use of the adverse-effect-ratio statistics is to model each component of the constituent of interest as a separate water-quality constituent in SELDM. For example, if the adverse effects of copper toxicity were a concern, then statistics for measured dissolved (filtered) concentrations could be used rather than, or in addition to, the whole-water concentrations of copper. Similarly, if the adverse effects of fine sediment in the water column were a concern, then statistics for measured concentrations of fine sediment could be used. If sediment deposition on the streambed were an issue, then statistics for measured concentrations of course sediment could be used. However, use of the adverse-effect ratios to model the concentrations of concern by using whole-water concentrations has at least six important advantages over the alternate approach for many constituents:

- Data needs—Modeling individual components of a whole-water constituent requires data to calculate the statistics to characterize upstream stormflow quality, highway-runoff quality, and BMP performance for each component.

- Storm-by-storm mass balance—The results from SELDM cannot be used to compare total concentrations with component concentrations on a storm-by-storm basis unless adverse-effect ratios or dependent water-quality relationships are used. Although SELDM will produce an output-value population that represents input statistics for each constituent definition, these constituent concentrations will be generated independently unless a dependent water-quality relationship is defined. The use of adverse-effect ratios, however, will ensure that the component of interest is some fraction of the total concentration modeled for each storm event.

- Data-quality concerns—Processing water samples to isolate the components of a water-quality constituent increases the probability of introducing physical and chemical bias in the results of the analyses (Patterson and Settle, 1976; Shiller and Boyle, 1987; Windom and others, 1991; Horowitz and others, 1992; Benoit, 1994; Benoit and others, 1997; Breault and Granato, 2003; Lopes and Dionne, 2003). For example, Benoit and others (1997) showed that ultraclean monitoring methods and materials were necessary to prevent order-of-magnitude increases in some dissolved trace element concentrations measured in natural water samples. Horowitz and others (1992) demonstrated that the measured concentration of dissolved (filtered) metal was dependent on the pore size, type, and diameter of the filter; the filtration method; and the volume of sample processed. The characteristics of the sample, including suspended sediment concentration, suspended sediment grain-size distribution, concentration of colloids and colloid-associated trace elements, and concentration of organic matter, also affected the proportion of the trace element measured as dissolved in these controlled experiments. Methods used to separate sediment grain-size fractions for analysis also may introduce bias and variability in reported concentrations. Thus, use of statistics for whole-water concentrations may be more robust than use of constituent fractions because of the potential effects of sample-processing artifacts.

- Detection-limit issues—The proportion of qualified values (most commonly values below detection limits) is commonly lower for the total concentration of many constituents than for the concentrations of the dissolved (filtered) components. For example, the number of qualified values in the filtered fractions of data for trace metal samples in the International

BMP Database is 1.12, 1.28, 1.71, 2.03, 1.87, 1.26, and 1.96 times the number of qualified values in the unfiltered fractions for Cd, Cr, Cu, Fe, Pb, Ni, and Zn, respectively. In the HRDB, the number of qualified values in the filtered fractions of data for trace metal samples is 2.61, 3.47, 0.392, 30.9, 6.69, 3.38 and 3.22 times the number of qualified values in the same sequence of unfiltered trace-metal fractions. Increasing the proportion of qualified values complicates analysis of water-quality data, increases uncertainties in calculated input statistics, and therefore increases uncertainties in model results (Helsel and Hirsch, 2002; Helsel, 2004; Granato and Cazenas, 2009). Thus, the use of statistics for whole-water concentrations may be more robust than the use of constituent fractions because of the potential effects of qualified data on sample statistics.

- Data availability—Data for the total concentrations of many constituents commonly are more plentiful than data for individual components. For example, Turcios and others (2010) found half as many sediment concentrations separately defined for sand or fines as the number of total suspended sediment concentration measurements in the USGS National Sediment Database. Similarly, queries to the July 2011 version of the International BMP Database indicated that the ratios of unfiltered to filtered analyses measured in BMP inflows and outflows were about 1.96, 1.28, 1.71, 2.03, 1.87, 1.26 and 1.96 for Cd, Cr, Cu, Fe, Pb, Ni, and Zn, respectively. Queries to the 2010 version of the HRDB (Smith and Granato, 2010) generated similar results with the ratios of unfiltered to filtered analyses measured in highway runoff of about 1.5, 1.45, 2.03, 4.7, 2.1, 1.22, and 1.96 for the same sequence of trace metals, respectively. For nutrients, the ratios of unfiltered to filtered concentrations of nitrite plus nitrate and phosphorus are about 26 and 8.5, respectively. Data-quality issues also limit the availability of data. For example, much of the dissolved trace element data collected by the USGS prior to the mid-1990s is in question (Windom and others, 1991; Smith and others, 1993). Similar limitations apply to most dissolved trace element, trace organic, and low-level nutrient data collected without use of clean sampling methods. Limits in available data reduce the opportunity to select input statistics based on data from hydrologically similar sites.

- Runoff-quality transformations—Modeling the dissolved fraction of a constituent or a single grain-size fraction of sediment is difficult because physical and chemical changes can occur in runoff. Sediment concentration and the geochemistry of runoff is expected to change as runoff moves over the pavement, through a drainage system, through a BMP structure, and into a receiving stream (Breault and Granato,

2003), especially if the receiving stream has sediment concentrations and geochemical characteristics that substantially differ from those in the highway runoff (Breault and Granato, 2003; Bricker, 2003). Use of the adverse-effects ratio, however, requires data or information about the fractionation of the constituents only in downstream receiving waters.

However, despite these limitations, modeling the concentration of concern as a separate water-quality constituent may be warranted if the results from this approach are deemed to be more defensible than the use of the adverse-effect ratio approach for a given situation.

SELDM uses the triangular/trapezoidal family of distributions for generating a stochastic sample of adverse-effect ratios for each specified downstream stormwater constituent. This family of distributions was selected for modeling the adverse-effect ratios because it can be parameterized by using expert judgment or by fitting the distribution to data (Johnson, 1997; Back and others, 2000; U.S. Environmental Protection Agency, 2001; Scherer, 2003; Kacker and Lawrence, 2007). The adverse-effect ratios must be greater than or equal to 0 and less than or equal to 1. SELDM uses the adverse-effects ratios rather than a chemical or physical constituent-speciation modeling module to estimate the concentration of concern. SELDM does not include a speciation-modeling module because different constituents and the associated concentrations of concern may be governed by different processes and, therefore, different variables.

The use of adverse-effect ratios, however, does require enough data or information to calculate statistics or to inform expert judgment. Results from chemical or physical constituent-speciation models may be used to estimate the parameters of the triangular/trapezoidal distribution of the adverse-effect ratios. For sediment, the grain-size distribution of bed sediments near highway outfalls and estimates of the grain-size distribution of the receiving waters will provide estimates of the concentration of concern. For nutrients, metals, and organic chemicals, the distribution coefficients from the literature can be used to estimate the proportions of constituent fractions in the receiving stream (U.S. Environmental Protection Agency, 1985a; Thomann and Mueller, 1987; U.S. Environmental Protection Agency, 1996c; Young and others, 1996; Tipping and others, 1998; Konstantinos, 2001; Allison and Allison, 2005). For example, Allison and Allison (2005) used published coefficient values and geochemical-modeling techniques to estimate statistics for the distribution coefficients of trace elements. Adverse-effect ratio statistics could be calculated on the basis of the values developed by Allison and Allison (2005), and the statistics for downstream suspended sediment concentrations could be calculated by using results from a preliminary run of SELDM for the site of interest. Geochemical-speciation models also could be used to estimate the concentrations of concern (Allison and others, 1990; Tipping, 1994; Allen and Hansen, 1996; Young and others, 1996; Di Toro and others,

2001; Bricker, 2003; Allison and Allison, 2005). SELDM can be used to calculate downstream water-quality statistics for pH and the concentrations of sediment, dissolved organic compounds, major ions, and other data needed to run such models. Although the concentrations for different constituents are randomly generated from storm to storm in SELDM, population statistics for each constituent can be used to represent a range of conditions for selective use as input to the geochemical models. Alternatively, for regulatory applications of SELDM, the adverse-effect ratio statistics could be estimated as part of an expert-judgment process that includes input from regulators and other stakeholders.

Mixing

SELDM calculates the concentrations, flows, and loads for the highway discharge (either direct runoff or BMP discharge), the upstream stormflow, and the downstream stormflow with the assumption that the downstream stormflow components are rapidly mixed (fig. 1). This is the standard assumption for almost all stormwater-quality models, including the 1990 FHWA model and the complex watershed models supported by the USEPA Basins program (Di Toro, 1984; Driscoll and others, 1989; Driscoll and others, 1990a, b; Adams and Papa, 2000; Zoppou, 2001; Walker, 2007; Rossman, 2010; Kuzin and Adams, 2010; U.S. Environmental Protection Agency, 2011). For example, Zoppou (2001) reviewed eight watershed models commonly used to evaluate the effects of stormwater runoff. He found that only one of these complex models included the advective-diffusion equations necessary for calculating mixing in the receiving stream. The output from SELDM provides bounding conditions for evaluating receiving-water quality. The highway-discharge concentrations represent the most conservative estimate of the risk for water-quality excursions. The fully mixed downstream concentrations, however, represent more realistic estimates of the risk for instream water-quality excursions; the estimates are somewhat conservative because the mass-balance calculations do not account for the potential effects of losses caused by physical, chemical, or biological attenuation; longitudinal dispersion; or downstream-flow accretion. These losses are expected to reduce instream concentrations of many constituents of potential concern as a pulse of stormwater moves downstream (Athayde and others, 1983; U.S. Environmental Protection Agency, 1985a, b, 1987; Thomann and Mueller, 1987).

The fully mixed runoff-quality approach was developed by the USEPA during NURP because the intermittent occurrence of runoff is expected to reduce adverse effects of excursions in comparison to steady-state discharges such as industrial or wastewater outfalls, and because the spatial and temporal distribution of runoff to receiving waters during storm events is complex (Athayde and others, 1983). For example, NURP evaluated the effects of short-term exposures that would result from intermittent stormwater runoff and estimated that acute (CMC) standards for runoff concentrations could be about twice those for steady-flow conditions while providing a similar degree of protection (Athayde and others, 1983). The intermittent occurrence of stormflow discharges can be assessed by using the precipitation statistics in table 2. If the highway-runoff duration is assumed to be about equal to the storm-event duration, then the average storm-event duration can be multiplied by the average number of runoff-producing events per year and divided by the number of hours in a year (about 8,766) to estimate the percentage of time during which highway runoff is discharging to the stream. On average, the duration of runoff flows would range from about 1.5 to 8.9 percent of the year in different rain zones under these assumptions. Alternatively, if it is assumed that a structural BMP is employed to extend the highway-runoff duration to 24 hours for each storm, then, on average, the duration of runoff flows would range from about 4.7 to 17 percent of the year in different rain zones.

The duration of runoff-producing events also can be used to examine the conceptual use of a mixing zone for stormwater discharges. If stormflow velocities in natural channels are on the order of 1 ft/s in low-slope basins (Chow and others, 1988), and the average runoff duration is on the order of 6 to 12 hours (table 2), then the pulse of runoff from the highway during such a storm would be distributed over a stream length of about 22,000 to 43,000 ft. These distances are expected to be well beyond the estimated hydrologic mixing zone for many streams of potential concern (Day, 1977; Fischer and others, 1979; Heard and others, 2001; Jeon and others, 2007; Divine and others, 2007; Gualtieri and Mucherino, 2008; Dow and others, 2009).

The fully mixed runoff-quality approach should meet DQOs for runoff-quality analyses because the uncertainties in runoff monitoring and modeling are large, and sophisticated hydrologic mixing-zone models do not represent conditions during storm events (Athayde and others, 1983; U.S. Environmental Protection Agency, 1985a). Mixing-zone analysis methods commonly do not include the complexities necessary to address wet-weather flows because these methods were largely developed for continuous point sources such as wastewater-treatment plants and industrial sources under steady-state extreme low-flow conditions (Fischer and others, 1979; U.S. Environmental Protection Agency, 1991; Dupuis, 2002; Divine and others, 2007; Oregon State Department of Environmental Quality, 2007; McCorquodale, 2007; Washington State Department of Ecology, 2010).

The USEPA and many states recognize the need to develop wet-weather criteria because highway and urban runoff commonly discharges during highly variable stormflow conditions. However, formal criteria have not been developed to date (Dupuis, 2002; Rachel Herbert, U.S. Environmental Protection Agency, Office of Water, Municipal Stormwater Program, written commun., 2012). Furthermore, mixing-zone equations may not represent conditions in small streams in which highway and urban runoff may be a substantial proportion of the downstream flows. Most mixing-zone studies and analyses have been limited to large low-gradient rivers or laboratory conditions that simulate conditions in such rivers (Day, 1977; Fischer and others, 1979; Heard and others, 2001; Jeon and others, 2007; Divine and others, 2007). Mixing-zone analyses commonly are based on assumptions that include simplified (far-field) flow distributions, minimal interaction of flow with irregularities on the streambeds and banks, and steady flow conditions without lateral inflows. Changes in slope and irregularities in streambeds, banks, and cross sections can accelerate transverse mixing so that the stream is rapidly mixed within about 2–25 stream widths (Day, 1977; Heard and others, 2001; Divine and others, 2007). Also, the rate of transverse diffusion increases substantially with increasing flows even with steady uniform flow in a clear, straight channel (Cotton and West, 1980). The hydraulic variables used to estimate mixing lengths also change with streamflow, distance along the stream, and time elapsed between large storms (Mackey, 1998; Heard and others, 2001).

The fully mixed runoff-quality approach also should meet DQOs for runoff-quality analyses because, unlike commonly used mixing-analysis methods, the fully mixed approach accounts for the characteristics of stormflow quality. Mixing models commonly are formulated with the assumption that background concentrations are negligible, and the tracer studies used to validate such models are designed so that background concentrations are negligible. For example, users of the moderately complex mixing model CORMIX are instructed to subtract the background concentrations from the effluent concentrations and the regulatory target concentrations (Jirka and others, 1996). This assumption may be warranted for wastewater discharges under steady-state low-flow conditions, but concentrations of constituents in stormwater from the upstream basin in many areas may be comparable to concentrations in highway runoff. As concentrations in the upstream stormwater and highway runoff converge, the effects of mixing are less distinct.

If, however, mixing zones are an issue for a given runoff study, then there are two types of mixing zones to consider, the regulatory mixing zone and the hydrologic mixing zone. The regulatory mixing zone is a limited area of the stream within which water-quality criteria may be exceeded. The USEPA has delegated the definition and regulation of constituent concentrations in mixing zones to the states, which use number of different approaches and definitions for assessing potential effects of effluents on receiving waters (U.S. Environmental Protection Agency, 1991; Dupuis, 2002; Divine and others, 2007; McCorquodale, 2007; Washington State Department of Ecology, 2010). The assumption is that the overall aquatic resource of the receiving water will be protected despite water-quality excursions within mixing zones because they are limited in size. For stormwater applications, such excursions also would be limited in duration. Different states may define the point of interest for regulating constituent-concentration excursions in the discharge itself (end-of-pipe concentrations), in a small area near the outfall (commonly known as a zone of initial dilution), or in a defined area downstream of the outfall but above the zone of complete mixing. Some states that regulate end-of-pipe concentrations adjust the target to a final acute value (FAV), which is generally the acute criterion multiplied by two. This FAV approach is similar to the NURP approach for concentrations of runoff constituents in receiving waters (Athayde and others, 1983). Generally, mixing-zone guidance approaches provide for routine acute water-quality excursions within all or part of a mixing zone and chronic excursions within and near the edge of the zone. Mixing-zone approaches commonly define a minimum zone of passage within the stream cross section where acute excursions are not allowed because the contaminant is assumed to be repulsive to mobile aquatic life. Regulatory mixing zones commonly are established by a rule-of-thumb based approach using one or more measures of channel geometry (Colorado State Department of Public Health and Environment, 2002; Oregon State Department of Environmental Quality, 2007; Washington State Department of Ecology, 2010).

The hydrologic mixing zone is the area in the stream between the discharge location and the location where the stream water is fully mixed with the discharge. Analysis of hydrologic mixing zones commonly centers on two processes, near-field and far-field mixing (Fischer and others, 1979; McCorquodale, 2007; Washington State Department of Ecology, 2010). Near-field mixing is the rapid and irreversible turbulent mixing of the water discharging from an outfall with the receiving water around the point of discharge. Near-field mixing is primarily caused by the momentum-induced velocity of the effluent and differences in the densities of the effluent and receiving waters. Near-field mixing characterization may help define zones of initial dilution. McCorquodale (2007) provides a summary of methods and models for estimating near-field mixing. Far-field mixing is generally caused by advection and dispersion and is generally much slower than the initial near-field mixing. Far-field-mixing analyses commonly are done to estimate the extent of regulatory mixing zones. If a mixing analysis is necessary, the development of the far-field plume can be estimated by using analytical equations (for example, equations by Fischer and others, 1979), simple models based on analytical equations (for example, FARFIELD and RIVPLUM5 developed by the Washington State Department of Ecology, 2010), or more complex models (for example, CORMIX developed by Jirka and others, 1996; or Visual Plumes developed by Frick and others, 2001). If analytical models are to be used, then it may be necessary to incorporate substantial adjustments to the transverse mixing coefficients published by Fischer and others (1979) to reflect advances in quantifying this variable for different instream conditions (for example, Cotton and West, 1980; U.S. Environmental Protection Agency, 1985a; Jeon and others, 2007; Divine and others, 2007; Gualtieri and Mucherino, 2008).

If the runoff-quality approach based on the assumption of complete mixing is deemed insufficient for a given application, then two conceptual modeling methods also can be applied by using SELDM to evaluate the risks for water-

quality excursions. The first method is upstream-flow reduction analysis, in which the SELDM analysis is copied (by using the copy option on the analysis form) and the upstream drainage area is reduced in proportion to the amount of stormflow in the regulatory mixing zone (William Fletcher, Oregon Department of Transportation, written commun., March 2012). The second method is the downstream flow-accretion analysis, in which the SELDM analysis is copied and the upstream drainage area is increased to account for the additional dilution caused by diffuse inflows of stormwater along the length of a stream even in the absence of a tributary stream or additional storm outfall. The downstream flow-accretion analysis also may be useful if there is a sensitive receptor, such as a water-supply intake, some distance downstream of the highway outfall to estimate the effect of any additional dilution between the outfall and the sensitive receptor.

The upstream flow-reduction analysis method is designed to assess additional risk for excursions in the regulatory mixing zone in comparison to the assumption that the inflow and the stream water are fully mixed. For example, in Oregon, the State Department of Environmental Quality (2007) established the width of a regulatory mixing zone to be less than 25 percent of the cross-sectional area of a river or stream. In this case, the risk of water-quality excursions could be assessed in the SELDM analysis by reducing the upstream drainage area to 25 percent of the actual value to proportionally reduce upstream stormflow volumes. However, the other hydraulic variables (drainage length, mean basin slope, impervious fraction, and BDF) and the hydrograph-recession statistics should not be modified to ensure that the timing of runoff from the upstream basin is not altered. Although it may be a concern, the population of transport-curve concentrations will remain the same because they were developed for normalized flow (in cubic feet per second per square mile) to scale stormflows to the basin size (Granato and others, 2009). Because the downstream flow volume is the sum of the upstream volume and the highway-runoff (or BMP-discharge) volume, this upstream-flow reduction may not result in a proportional reduction in the dilution factor or a proportional increase in the risk for excursions in all storms; especially in small or arid basins, to which the highway may contribute a large proportion of the stormflows. The downstream water-quality and dilution-factor output from the upstream-flow reduction analysis with SELDM can be used in conjunction with the modeling results for the fully mixed condition to assess the potential for water-quality excursions downstream of the site of interest.

The downstream flow-accretion analysis method is designed to calculate the reduction in risk for water-quality excursions caused by additional dilution that is expected to occur in the hydrologic mixing zone downstream of an outfall (as defined by a far-field mixing analysis) during storm events. The ratio of the downstream area to the upstream area is the downstream flow-accretion factor that can be used to estimate dilution at the point of interest. Analyses of relations between basin length (defined as the length of the main channel from the outlet to the watershed divide) and drainage area show strong correlation for the 845 sites with drainage areas greater than 0.1 mi[2] in the secondary dataset compiled by Granato (2012) (fig. 26). Use maps, GIS, or Streamstats (Ries and others, 2008) to delineate the downstream basin, or add the estimated mixing length to the upstream basin length and use the equations in figure 26 to estimate the total downstream area. If there are tributaries or substantial stormwater outfalls that enter the stream within the estimated downstream mixing length, use the delineation method. Otherwise, the equations in figure 26 can be used to (1) calculate an equivalent length of the upstream basin at the highway outfall by using the (predetermined) upstream drainage area, (2) add the length of the mixing zone to this estimated basin length, (3) calculate the downstream drainage area by using the second regression equation with the total estimated basin length, and (4) calculate the flow-accretion factor by dividing the downstream by the upstream drainage areas. If figure 26 is used in this way, it is necessary to use the lengths and areas calculated by using the line of organic correlation on the graph, because (as evidenced by the scatter of data above and below the line) use of the actual upstream basin length plus the mixing length may result in a calculated downstream area that is less than the measured upstream area.

Once the flow-accretion ratio is calculated, this value can be used to assess the risk of water-quality excursions at the point downstream of the outfall by adjusting the estimated input drainage area in a SELDM analysis or the model output directly. To run the scenario in SELDM, the modeled upstream drainage area would be increased to proportionally increase stormflow volumes. As in the flow-reduction analysis, however, the other hydraulic variables (drainage length, mean basin slope, impervious fraction, and BDF) and the hydrograph-recession statistics should not be modified to ensure that the timing of runoff from the upstream basin is not altered. To adjust simulation results outside SELDM, apply the flow-accretion factor to the upstream flows to adjust the dilution factors and the fully mixed concentrations.

The flow-accretion factors in figure 27 demonstrate that the additional stormflow contributions that occur over relatively short mixing zones may be substantial, especially for small upstream drainage basins. For example, inflows from the additional contributing area within a 500-ft mixing zone downstream of a highway outfall would be expected to produce about 40 percent more dilution than is available at the highway outfall for a 0.1-mi[2] upstream basin and 10 percent more for a 1-mi[2] basin (if the quality of inflows is assumed to be comparable to the quality of upstream stormflows). The additional drainage area associated with a 2,500-ft mixing zone is expected to provide about 14 percent more dilution for a 10-mi[2] basin. The magnitude of lateral inflows calculated by using the flow-accretion curves in figure 27 also indicates the potential for turbulent transverse mixing caused by additional

stormflow outfalls or diffuse lateral inflows along the stream during storm events. The substantial accretion of lateral inflow volumes over short mixing distances as evidenced by these curves also indicates that steady state (dry-weather) mixing-zone models may substantially underrepresent dilution downstream of a stormwater outfall during wet-weather conditions.

Flow-accretion factors also may be used without a detailed mixing-zone analysis. For example, if the quality of water at a water-supply intake 1 mi downstream of a highway outfall is of concern, then the fully mixed downstream concentrations at the intake could be estimated by calculating the flow-accretion factor. If the drainage area of the upstream basin at the highway outfall is 1 mi[2], then the potential downstream dilution at the intake would be about 2.2 times the potential dilution at the outfall; if the area of the upstream basin at the highway outfall is 10 mi[2], then the potential downstream dilution at the intake would be about 1.4 times the potential dilution at the outfall (figs. 26, 27). If the duration of the runoff pulse from a highway or BMP during the average storm is considered, the flow accretion over the 20,000- to 40,000-ft length of the pulse in the absence of large downstream tributaries would be about 8 to 20 times the potential dilution at the outfall if the drainage area of the upstream basin at the outfall is 1 mi[2] and about 2 to 4 times the potential dilution at the outfall if the drainage area of the upstream basin at the outfall is 10 mi[2] (fig. 27). The concentrations calculated by assuming instantaneous mixing at the outfall may be considered a conservative estimate of the risk for water-quality excursions at downstream locations because of the additional dilution and mixing of runoff from contributing areas between the outfall and the downstream point of interest.

Lake-Basin Analysis

The SELDM lake-basin module produces a population of average annual constituent concentrations in a lake that receives highway runoff. This component of SELDM is designed to provide planning-level estimates for an assessment of the potential effects of highway runoff on the water quality in a downstream lake or pond. This lake-basin module, a lumped-parameter model, treats the highway, the lake basin, and the lake as simple control volumes rather than detailed, spatially distributed features. This lake-basin module is a simple implementation of the Vollenweider mass-balance model, which is well established in the literature (Vollenweider, 1975; Reckhow, 1979; U.S. Environmental Protection Agency, 1983, 1985a, c, 1986a; Driscoll, 1990b; Thomann and Mueller, 1997). The Vollenweider mass-balance model is commonly used for planning-level estimates of annual average lake concentrations for TMDL studies (U.S. Environmental Protection Agency, 2000, 2007b) and is based on eight basic assumptions:

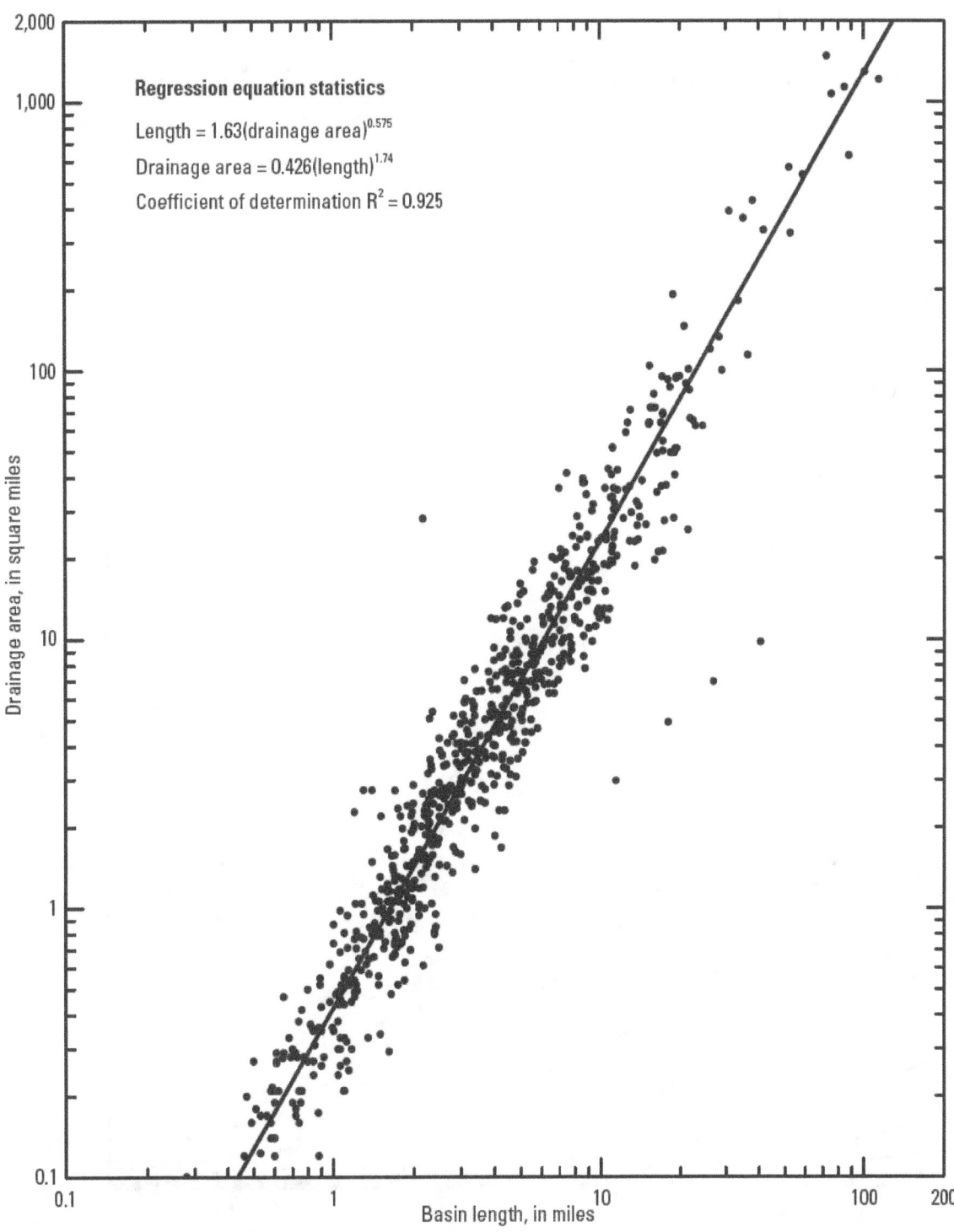

Figure 26. Relations between the basin length, defined as the length of the main channel in miles from the outlet to the drainage divide, and the basin drainage area in square miles for 845 sites that have drainage areas greater than 0.1 mi² in the secondary dataset compiled by Granato (2012). The regression equations were calculated by using the line of organic correlation because it provides a unique line for using either variable as the explanatory variable (Helsel and Hirsch, 2002).

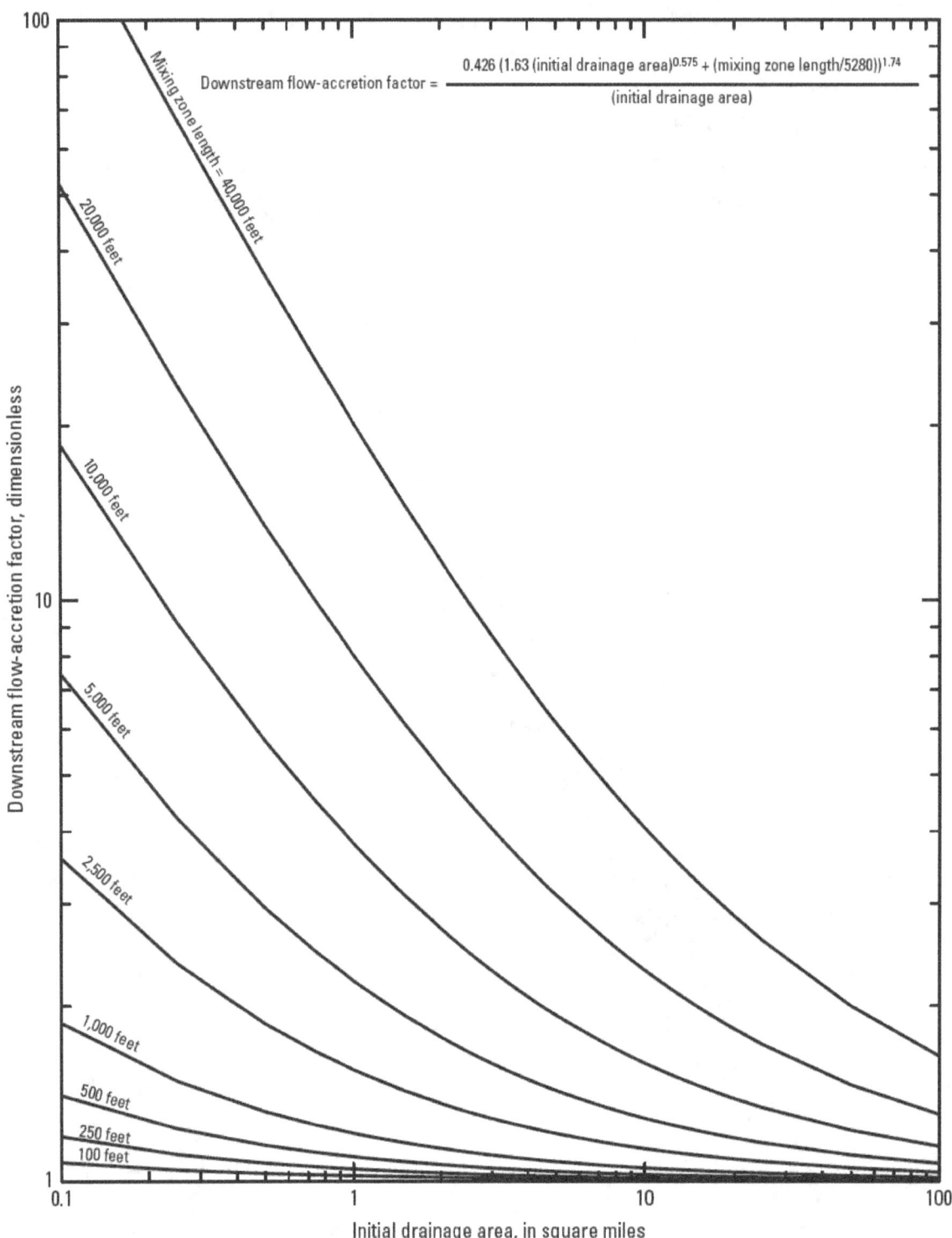

$$\text{Downstream flow-accretion factor} = \frac{0.426\ (1.63\ (\text{initial drainage area})^{0.575} + (\text{mixing zone length}/5280))^{1.74}}{(\text{initial drainage area})}$$

Figure 27.   The downstream flow-accretion factors, which are the ratios of downstream flows at the ends of the specified mixing-zone lengths to the flows immediately downstream from the highway outfall.

- the lake is a fully mixed control volume;

- storage does not change on an annual basis so inflows equal outflows;

- regional streamflow statistics, which include areas of groundwater recharge, discharge, and lakes, are representative of the total flux of water through the lake on an annual basis;

- each year in the analysis represents an independent "event" for the purpose of calculating annual average concentrations;

- annual runoff volumes and loads from the drainage areas to one or more highways in a basin can be modeled as output from a single site representing the combined area for all highway sites;

- annual runoff volumes and loads from the rest of the basin can be modeled as output from a single area representing the total lake-basin area minus the drainage areas of the highway(s);

- bed sediments are not a substantial source of solids or water-quality constituents in the part of the water column being modeled to estimate the average concentration; and

- reductions in concentrations of each water-quality constituent in the lake can be modeled as a single attenuation factor that represents applicable first-order decay constants.

Without such assumptions, the lake-modeling effort would be complex and require detailed spatial and temporal data and the use of a complex model such as the Water Quality Analysis Simulation Program (WASP), which is developed and maintained by the U.S. Environmental Protection Agency (2010). In addition, the results of a complex modeling effort may not be substantially better than a planning-level estimate without a long and detailed set of calibration data (Granato, 2010). For example, Bhavsar and others (2008) compared results from two models that included coupled fate and transport of metals in lakes and found that, despite a complex modeling effort with detailed characterization data, simulated concentrations differed from measured concentrations by as much as 400 percent. Bhavsar and others (2008) indicated that the results were constrained by a lack of detailed calibration data rather than the structure of the models.

Mass Balance Approach

The lake analysis is based on two independent components for annual lake output—the highway-runoff component and the lake-basin component—for all downstream water-quality pairs identified as part of the lake analysis. A process-flow diagram is shown on figure 28. The highway-runoff-load component is generated storm-by-storm in the

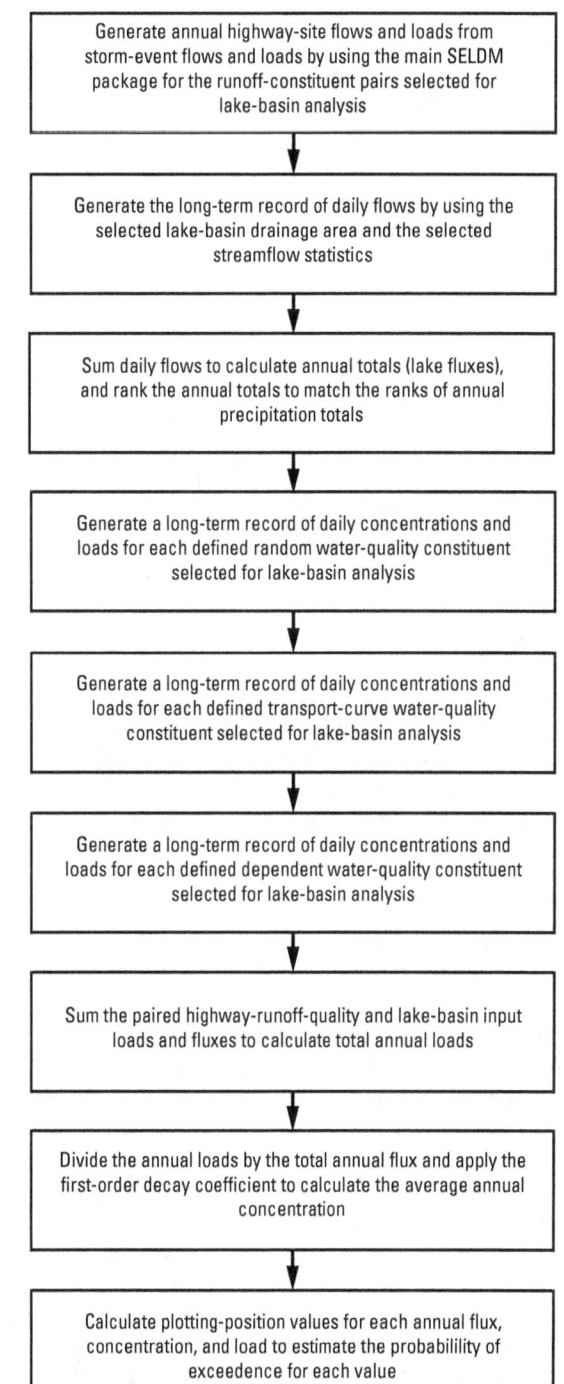

Figure 28. The steps used by the Stochastic Empirical Loading and Dilution Model to estimate lake-basin fluxes, loads, and concentrations.

main SELDM module. The lake-basin flows, concentrations, and loads are generated for each day during the entire period, aggregated into annual sums, and paired with the highway-runoff component. The selected streamflow statistics, which are adjusted for the entire lake-basin drainage area, are used to generate the daily lake-basin flow values. Individual daily values generated for the lake-basin analysis should not be exported and used for further analysis because SELDM generates each value randomly without serial correlation or seasonal patterns; the annual-load accounting years are just random collections of 365 or 366 sequential daily values. Lake-flux years are assigned to highway-runoff-analysis years by pairing values that are ranked respectively by total annual streamflow volume and total annual precipitation volume. The annual average steady-state concentrations are estimated from measures of lake volume, annual loads, flux through the lake, and an attenuation factor (which may include physical settling and other attenuation mechanisms).

The properties of the lake, including drainage area, surface area, and mean depth, are entered on the lake-basin form in SELDM. The lake-basin drainage area (in square miles) is the entire contributing area to the lake upstream of the lake outfall (if there is a surface-water outfall) minus the area of the highway. The lake-basin drainage area is used to calculate the total annual flux of water through the lake (exclusive of highway runoff) based on regional, local, or site-specific streamflow data. In most cases, the surface-water divide of the lake basin may be used to approximate the contributing area to the lake. In some cases, however, streamflow data or groundwater maps may indicate that the surface-water and groundwater divides do not coincide. In such cases, the analyst may adjust the streamflow statistics or the lake-basin drainage area to best represent the total annual flux through the lake-aquifer system. The highway drainage area (in acres) can be determined for a given stretch of highway or for the total length of all highways in the basin. If runoff from the highway site discharges into a single tributary within a lake basin, then the fluxes through the stream analysis and the lake analysis may be done together by using a physical basin-delineation method, shown schematically in figure 29A. In this case, the total lake-basin area is the sum of the upstream and downstream basin areas, including the lake, but not the highway. If, however, a highway or highways intersect multiple tributaries, then the total highway area can be modeled conceptually as a single contributing area and the lake-basin area as the total watershed area minus the area of the modeled highways (fig. 29B). In this case, the SELDM output for downstream concentrations for individual storm events would not represent conditions in any of the individual tributaries because the highway and upstream areas do not correspond with the drainage areas at any particular stream crossing. In either case, the highway contributions are entirely from storm-event runoff, and the lake-basin contributions are from total annual runoff (base flow plus stormflow).

The SELDM lake-basin model is a fully mixed control volume because it does not account for variations in vertical or spatial concentrations within the lake. The surface area of the lake, in acres, is defined as the annual mean surface area, which can be estimated from topographical maps, a geographic information system, or other sources of information. The surface area is multiplied by the mean depth to estimate the total volume of the lake. The total volume of the lake is used with the total annual flux of water through the lake to estimate the hydraulic residence time of the lake. On an annual basis, this approach is suitable for lakes with large surface-area-to-depth ratios with many small inlets and may be used as a simplified approximation for many lakes (U.S. Environmental Protection Agency, 1985c, 1986a). If a lake is stratified, this approach may be used to approximate conditions in the active layer of the lake if the modeled volume and depth represent the volume and depth of this layer. Defined as such, the hydraulic residence time is defined as the average time for any given year during which a water molecule stays in the lake rather than the time required to exchange all water in the lake:

$$T_w = \frac{V}{Q},\qquad(24)$$

where

T_w is the mean hydraulic residence time of water, in years;

V is the volume of the lake in cubic feet; and

Q is the annual flux through the lake, in cubic feet per year.

If it is assumed that there is no change in storage from year to year, then the lake volume (V) is modeled as the surface area times the mean depth (a constant), and the annual mean residence time of water in the lake (T_w) is the parameter that changes with variations in annual flux (Q) so that annual inflows will equal outflows.

Estimates of the mean depth, however, are not as readily available as of the surface area of the lake. Bathymetric maps commonly are available from Federal and state land and water-management agencies (for example, Guthrie and Stolgitis, 1977; New Jersey Division of Fish and Wildlife, 2008; Connecticut Department of Environmental Protection, 2010; Massachusetts Division of Fisheries and Wildlife, 2010; Rhode Island Department of Environmental Management, 2010; University of Florida Lakewatch, 2010). Water-supply agencies may have compiled bathymetric maps for active and inactive water-supply reservoirs. For example, Waldron and Archfield (2006) provide lake stage, volume, and surface area data and relations for 44 reservoirs in Massachusetts as part of a water-supply analysis. Currently (2012), instruments that combine global positioning system and sonar sensors have been integrated for use by recreational boaters. Such systems, when coupled with GIS or mapping software, are able to provide mean depths that have a high degree of accuracy. An initial estimate of average depth can be made by using a modification to a regression equation developed by Bartsch

A. Physical lake-basin example

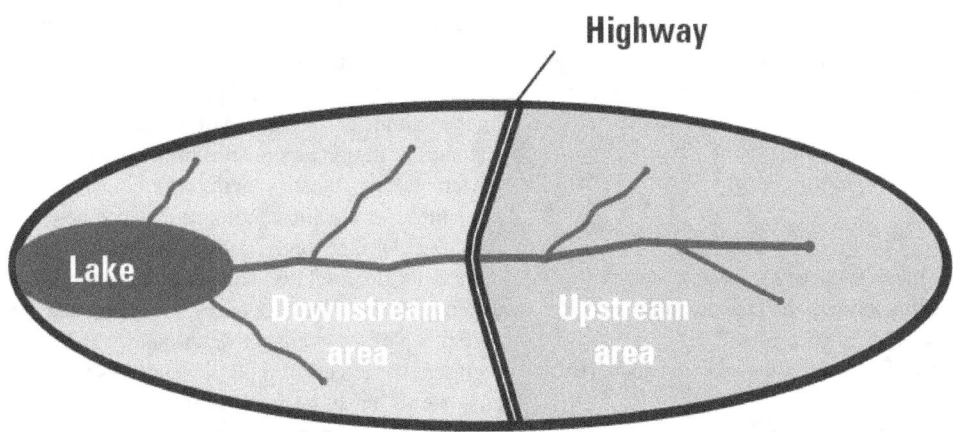

Lake basin area = downstream area + upstream area
= total basin area - highway area

B. Hypothetical lake-basin example

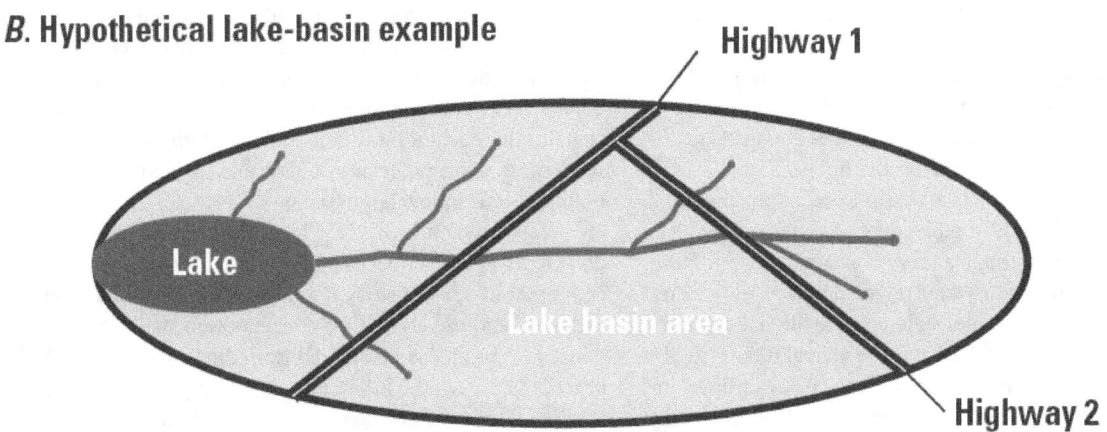

Lake basin area = total basin area - highway area
Highway area = highway 1 + highway 2

Figure 29. Examples of lake-basin areas defined by *A*, a physical delineation (used when there is a clear distinction between areas that are upstream and downstream of the highway) and *B*, a hypothetical delineation (used if the highway area does not divide the basin, or if there are multiple sections of highway).

and Gakstatter (1978) for predicting the mean hydraulic residence time in days from the ratio of drainage area to surface area for 36 lakes and reservoirs in the northern United States. If the mean hydraulic residence time is expressed as in equation 24, and the volume is divided by the surface area of the lake, the equation developed by Bartsch and Gakstatter (1978) becomes

$$D_L = 32.7 \times Q \times SA^{0.177} \times DA^{-1.177} , \qquad (25)$$

where

D_L	is the mean depth of the lake, in feet;
Q	is the flux through the lake, in cubic feet per year;
SA	is the surface area of the lake, in square feet; and
DA	is the drainage area of the lake, in square feet.

Equation 25 was tested for nine lakes in Massachusetts with basin areas, surface areas, and mean depths that had been documented for TMDL analyses. The error in resulting predictions ranged from -40 to 180 percent with a median error of 8 percent and a median absolute deviation of 41 percent. For this reason, estimates based on equation 25 should only be used for an initial rough planning-level estimate.

The SELDM lake-basin module uses regional streamflow statistics to estimate the surface-water and groundwater flux through the lake (Q) in each model year. These streamflow statistics are based on data from streamgages downstream from areas of groundwater recharge and discharge and lakes; thus, streamflow data is assumed to be an acceptable surrogate for modeling the total amount of surface water and groundwater flowing through a lake on an annual basis. This assumption is commonly used for regional hydrologic studies (Granato, 2010). For example, regional and national contour maps produced by the USGS of average annual runoff (total streamflow) provide contours of equal annual streamflows in inches, which are equivalent to the long-term average of daily mean streamflows (Langbein, 1949; Busby, 1966; Gebert and others, 1987; Randall, 1996). The FHWA used streamflow estimates that were generalized from a national annual-runoff map to estimate the effects of highway runoff on streams and lakes (Driscoll and others, 1990a, b). The USEPA also has developed runoff-contour maps for use in planning-level analyses of water-quality data (Bishop and Church, 1995).

However, annual flux estimates based on normalized streamflow statistics can be inaccurate for areas where surface and groundwater divides are substantially different if the total contributing area is not included. For example, Granato and others (2003) found a large difference in streamflows caused by differences in groundwater and surface water divides in a simulation of the hydrogeology of the Big River Area, Rhode Island. In this study, the surface-water drainage area of the Mishnock River at Route 3 was measured as 0.29 mi², and streamflow measurements indicated that the long-term average flow would be about

14.9 ft³/s/mi². However, the simulation indicated that the groundwater contributing area to Lake Mishnock was 1.6 mi². The normalized average flow from the Lake Mishnock groundwater contributing area was about 2.7 ft³/s/mi², which is much closer to the average for nearby streamgages.

The SELDM user can specify streamflow statistics based on the latitude and longitude of the highway site and regional statistics or designate statistics from nearby streamgages on the streamflow-statistics input form. User-defined statistics for modeling site-specific conditions can also be entered on the form. SELDM, however, does not currently include the option to model different flows, concentrations, and loads from different areas of the lake basin. This approach is consistent with much of the available water-quality and flow data, which represent the effects of multiple land uses instead of only one land use within an upstream drainage area, and is evidenced by the large drainage areas of sampling sites associated with available water-quality data. For example, Granato and others (2009) obtained data from 24,581 USGS surface-water-quality-monitoring stations in the conterminous United States and found that the median upstream area to sites where data for total nitrogen, total phosphorus, total hardness, and suspended sediment concentrations had been collected was about 50 mi² and ranged from about 0.01 to 1,000 mi². Fewer than 25 percent of these basins had drainage areas less than 10 mi². Examination of the available surface-water-quality-monitoring data for pH, nitrate plus nitrite, chloride, total copper, total lead, and total zinc showed similar drainage-area statistics for each of these constituents. With the exception of wilderness areas, rangeland, and farmland, land uses would be homogeneous in few large basins.

If, however, the SELDM lake-basin analysis is being done to assess the effect of a highway on lake-water quality in support of a wider study that includes load estimates for different land-use areas, then concentration statistics for the entire lake basin can be derived from the loads assigned to the other land uses in the basin. For a planning-level analysis, the average flow-weighted concentration from all nonhighway land uses in the basin is the sum of the annual loads divided by the sum of the annual flows from each area. The variability of the background water-quality component can be estimated by using published values of the COV for different water-quality constituents (for example, Athayde and others, 1983; Pitt and Vorhees, 2003). The NURP values for the COV of TSS, chemical oxygen demand (COD), total phosphorus, total copper, total lead, and total zinc range from 0.85 to 2.92, 0.39 to 0.78, 0.69 to1.66, 0.81 to 1.32, 0.68 to 1.52, and 0.66 to 1.07, respectively (Athayde and others, 1983). In the NURP study, the least developed land uses were associated with the highest COV values, and the COV values for urban land uses were similar for most constituents. Once the average and COV of flow-weighted concentrations have been selected from the literature or from available data, the lognormal statistics to be used in SELDM can be calculated by using the standard methods (Chow, 1954; Driscoll and others, 1990b; Stedinger and others, 1993) described in appendix 1.

Under steady-state conditions, the sum of the input loads equals the sum of the output loads and the attenuated loads. If the attenuation factor is expressed as a first-order decay process, then the mass balance equation for the lake can be expressed as

$$Q_{in}C_{in} = Q_{out}C_{lake} + KC_{lake}V_{lake}, \qquad (26)$$

where

Q_{in} is the total of inflows to the lake, in units of volume per time;

C_{in} is the average inflow concentration, which is calculated by dividing the sum of the input loads by the sum of the input flows, in units of mass per volume;

Q_{out} is the total of outflows from the lake, in units of volume per time;

K is the attenuation factor, in units of the reciprocal of time;

C_{lake} is the average concentration in the lake, in units of mass per volume; and

V_{lake} is the volume of the lake

With the assumption of steady-state lake volume, Q_{in} is equal to Q_{out}, which is defined as the flux through the lake and denoted as Q_{flux}, equation 26 can be rearranged to solve for C_{lake}. To obtain the equation as a function of the mean residence time of water in the lake, the right side of the equation is divided by the flux through the lake:

$$C_{lake} = \frac{Q_{flux}C_{in}}{Q_{flux} + KV_{lake}} = \left(\frac{Q_{flux}C_{in}}{Q_{flux}}\right) \div \left(\frac{Q_{flux}}{Q_{flux}} + \frac{KV_{lake}}{Q_{flux}}\right), \quad (27)$$

and finally

$$C_{lake} = \frac{C_{in}}{1 + KT_W}, \qquad (28)$$

where

T_w is the mean residence time of water in the lake.

Equation 28 is the form of the equation used in many implementations of the Vollenweider model (U.S. Environmental Protection Agency, 1983, 1985c, 1986a; Thomann and Mueller, 1997) and for calculation of primary settling of sediment for water- and wastewater-treatment analyses.

Attenuation Factors

The water-quality parameters to be modeled in the SELDM lake module and the associated attenuation factors are selected on the water-quality constituent-pair specification form. The attenuation factor is a first-order rate coefficient defining the change in concentration with time. For sediment,

the attenuation factor is calculated by dividing the net effective settling velocity (in units of length per time) by the mean depth of the lake (in units of length). For other constituents, the attenuation factor can be modeled as a first-order decay rate, which is the sum of the attenuation factors for different processes including sedimentation, volatilization, biodegradation, photolysis, and hydrolysis (U.S. Environmental Protection Agency, 1985a, b, c; 1986a; 1987; Thomann and Mueller, 1987). The attenuation factor can be expressed as

$$K_T = K_S + K_V + K_B + K_P + K_H, \qquad (29)$$

where

K_T is the total attenuation factor;

K_S is the attenuation factor from particulate settling;

K_V is the total outflow from the lake, in units of volume per time;

K_B is the attenuation factor from biodegradation, which is the transformation of one constituent into others as a result of biological metabolic processes;

K_P is the attenuation factor from photolysis, which is the transformation of one constituent into others as a result of solar radiation; and

K_H is the attenuation factor from hydrolysis, which is the transformation of one constituent to others as a result of interaction with water molecules.

SELDM uses one attenuation factor for each defined constituent because the state of knowledge for specifying these numbers on an annual basis does not warrant a more complex approach. The user can specify multiple constituent pairs or rerun the analysis with different attenuation factors to do a sensitivity analysis for this variable. If data are available, the user also may model different fractions (for example, the dissolved or sediment-associated fraction or different grain-size fractions) as separate constituents with different attenuation factors. The user also may estimate the long-term average concentrations and attenuation factors for each fraction, calculate the resulting annual average concentration for each fraction, and then back calculate the long-term-average bulk attenuation factor from the sum of the component concentrations.

Nutrients

The published attenuation factors for nutrients are commonly accepted because the effects of nutrients on lakes have been studied extensively. For example, Thomann and Mueller (1987) report net effective settling velocities of 32 to 41 ft/yr and literature values ranging from about 2 to 656 ft/yr for total phosphorus. Canfield and Bachmann

(1981) report mean net effective settling velocities for total phosphorus of 79 ft/yr (with values ranging from -85 to 148 ft/yr) for 290 natural lakes and 407 ft/yr (with values ranging from -951 to 1,610 ft/yr) for 433 artificial lakes. Walker (1987) presents data from a literature review indicating that the net effective settling velocity for total phosphorus in seven urban lakes ranged from 29 to 162 ft/yr. Bachmann (1981) reports mean net effective settling velocities for total nitrogen of 16 ft/yr (with values ranging from -27 to 427 ft/yr) for 95 natural lakes and 29 ft/yr (with values ranging from -164 to 1,290 ft/yr) for 384 artificial lakes. Negative values of the mean net effective settling velocities may indicate net production of nutrients, conditions that were not steady state during the sampling period, or errors of estimation (Bachmann, 1981; Canfield and Bachmann, 1981). Researchers relate nutrient attenuation factors to loading rates, hydraulic residence time, mean depth, and other factors (Bachmann, 1981; Canfield and Bachmann, 1981).

Sediment

Attenuation factors for sediment are based on the effective settling velocity of suspended sediments. Estimating attenuation factors for suspended sediment is important because particulate settling is a large part of the total attenuation factor for many water-quality constituents. Commonly accepted attenuation factors for suspended sediments are not readily available because of large temporal and geographic variations in the concentrations and physical properties of sediment. Suspended sediment includes all particulate matter, including mineral grains, plant matter, small aquatic organisms such as bacteria, and trash. The attenuation factor used for the lake module is for the sediment that is transported to the lake and thus depends on many natural and human factors. Sampling, sample handling, and analysis methods can have a substantial effect on measured values of the concentration, density, and grain size of sediment in the water column (Guy, 1977; Edwards and Glysson, 1999; Bent and others, 2003). Information from the literature can guide initial estimates, but knowledge of local conditions is necessary to refine such estimates. For many areas, such data are available from the USGS NWISWeb database. Turcios and others (2010) cataloged the amount of grain-size data available for the United States in the USGS NWISWeb database as of January 2000 and found that tens of thousands of measurements had been collected at thousands of sites in the United States. The water-quality data-mining techniques and software developed by Granato and others (2009) can be used to obtain data on the concentration and grain size of sediment from the USGS NWISWeb database.

Attenuation factors for suspended sediment commonly are calculated from measurements of settling velocities of equivalent spheres in a quiescent fluid (U.S. Environmental Protection Agency, 1987; Thomann and Mueller, 1987). Hallermeier (1981) indicates that the settling velocities for spheres are a good approximation for the settling velocities of sand grains. Gibbs and others (1971) developed an equation based on the viscosity and density of water and the density of the spheres to calculate the settling velocities of spheres in a quiescent fluid (in centimeters per second). This equation is useful for the lake analysis because it is valid for the full range of sediment sizes, whereas the commonly used Stokes law diverges from experimental data under standard conditions for grain sizes larger than about 20 microns (a grain size in the range from median to coarse silt). When converted to feet per year for use in calculating the SELDM attenuation factor, the equation is

$$v_s = CF \frac{-3\eta + \sqrt{9\eta^2 + gr^2 P_f (P_s - P_f)(0.015476 + 0.19841r)}}{P_f (0.011607 + 0.14881r)}, \quad (30)$$

where

V_s is the settling velocity of a sphere, in feet per year (ft/yr);

η is the dynamic viscosity of the fluid, in poises (0.01002 at 20°C);

g is the acceleration caused by gravity, in centimeters per second per second (980.665 cm/s^2);

r is the sphere radius, in centimeters;

P_f is the density of the fluid, in grams per cubic centimeter (0.9997 g/cm^3 at 20°C);

P_s is the density of the sediment, in g/cm^3; and

CF is the conversion factor from centimeters per second to feet per year (1,035,354 s ft/cm yr).

Settling velocities calculated by use of the equation by Gibbs and others (1971) for spheres with different grain-size diameters and densities are shown in figure 30. The velocities range over nine orders of magnitude and are expressed in feet per year to facilitate calculation of annual attenuation factors with average lake depth from SELDM.

In comparison, Driscoll and others (1986) summarized settling velocities measured during NURP in five equal size classes by mass as about 263, 2,630, 13,100, 61,400, and 570,000 ft/yr, but they did not provide information on the grain-size or density distributions of these particles. Dorman and others (1996) measured settling velocities in highway runoff by the same methods used by Driscoll and others (1986) and found that highway-runoff sediments had a lower proportion of particles in the first four settling-velocity classes (about 17 percent of total mass in each class) and a larger proportion (about 28 percent of total mass) in the category with the highest settling velocity (570,000 ft/yr). Dorman and others (1996) also did not provide information on the grain-size or density distributions of these particles. Grain-size information is needed, however, to calculate size-specific attenuation factors or a single weighted-average attenuation factor. The grain-size diameters used to calculate fall velocities displayed on figure 30 range five orders of magnitude and are shown with corresponding sediment-size classes.

Studies of highway and urban runoff indicate large ranges in the masses associated with measured grain sizes in runoff sediments from storm to storm and site to site (Sansalone and Tribouillard, 1999; Li and others, 2006; Smith and Granato, 2010). For example, Sansalone and Tribouillard (1999) showed that about 0, 10, 60, and 30 percent of the sediment by mass in pavement runoff is in the clay, silt, sand, and gravel categories, respectively. Li and others (2006) collected 16 grab samples from 3 sites where runoff exited a drainage pipe; they did particle counts and measured the mass of 11 size fractions between 2 and 1,000 microns and found that more than 90 percent of the particles were less than 10 microns in diameter, but this fraction contributed only about 10 percent of the mass of particulates in runoff. About 5, 30, and 60 percent of the particulates in the samples described by Li and others (2006) were in the diameter ranges for clay, silt, and sand, respectively. Fowler (2008) studied 13 samples of parking-lot runoff and found that the median diameters for the 10th, 50th and 90th percentiles were 4, 50, and 1,400 microns, respectively. Kim and Sansalone (2008) showed grain-size distributions in runoff from seven studies, including their own, indicating that grain sizes in noncolloidal fractions range from 1 to larger than 24,500 microns with median grain sizes ranging from 20 to 800 microns. Kim and Sansalone (2008) also indicated that many studies of urban and highway runoff show similarities in grain-size distributions. Smith and Granato (2010) documented the percentages of the masses for three sediment-size fractions—less than 62.5 microns (clay and silt), 62.5 to 250 microns (very fine to medium sand), and greater than 250 microns (coarse sand and gravel)—in 162 highway-runoff samples from 15 sites in Massachusetts. They found that the average mass percentages for each fraction were 61, 15, and 24, respectively. The percentage ranges for the three size fractions were 4–99, 0–48, and 0–87 percent, respectively. Selbig and Bannerman (2011) provided detailed data for the particle-size distribution in 20–90 samples of runoff from a roof, three parking lots, a feeder street, a connector street, an arterial street, and a mixed land-use basin in the area around Madison, Wisconsin. They measured the percentages of the mass of particulates in 10 grain-size intervals from less than 2 to less than 500 microns.

Grain-size distributions in receiving waters vary substantially. Vice and others (1969) studied a basin undergoing development in Virginia and found that the percentages of clay, silt, and sand and gravel by mass during the 1961–64 period were 26, 60, and 14 percent, respectively. The percentages of clay, silt, and sand and gravel transported in stormflows were 25, 61, and 14 percent, respectively. The percentages of the masses of clay, silt, and sand and gravel transported in nonstorm flows were 74, 26, and 0 percent, respectively, but nonstorm sediment discharges accounted for less than 1 percent of the total sediment discharge during the study period. Similarly, Horowitz and others (2008) intensively monitored the water quality in 10 urban streams in Atlanta, Georgia, and found that stormflows accounted for more than 65 percent of annual streamflows and more than 94 percent of suspended sediment loads. Yorke and Herb (1978) measured the particle-size distribution in 19 medium- to high-flow samples collected during the period 1960–73 in a basin in Maryland. Sediment transport in this basin was being studied because of rapid suburban growth; development on about 3 percent of the undeveloped land surface each year during the study period resulted in an increase in total impervious area from about 4 to 9 percent of the watershed area. The mean percentages by mass of clay, silt, and sand and gravel measured in this study were 35, 45, and 20 percent, respectively. The percentage ranges of the clay, silt, and sand and gravel fractions were 11–73, 21–62, and 3–38, respectively. Yorke and others (1985) calculated loads of sediment in the Schuylkill River in Pennsylvania from 1954–79 and reported that suspended sediments comprised about 40, 54, and 6 percent of clay, silt, and sand, respectively. Slattery and Burt (1997) indicated that about 13, 65, and 22 percent of suspended sediments in a stream in an agricultural area comprised clay, silt, and sand, respectively. Holnbeck (2005) measured suspended sediment on the upper Yellowstone River in Montana and found that, on average, about 51 percent of suspended sediment was silt or clay, and the remaining 49 percent was sand; these percentages varied with streamflow.

Grain-size distributions in lake sediments indicate the characteristics of sediment in flows from contributing areas. These grain-size distributions, however, vary substantially both spatially and temporally (Gottschalk, 1961; Guy, 1970). Gottschalk (1961) indicated that the percentage ranges of the masses of clay, silt, and sand in reservoir deposits were 9–85, 14–78, and 0–77, respectively. The average percentages by

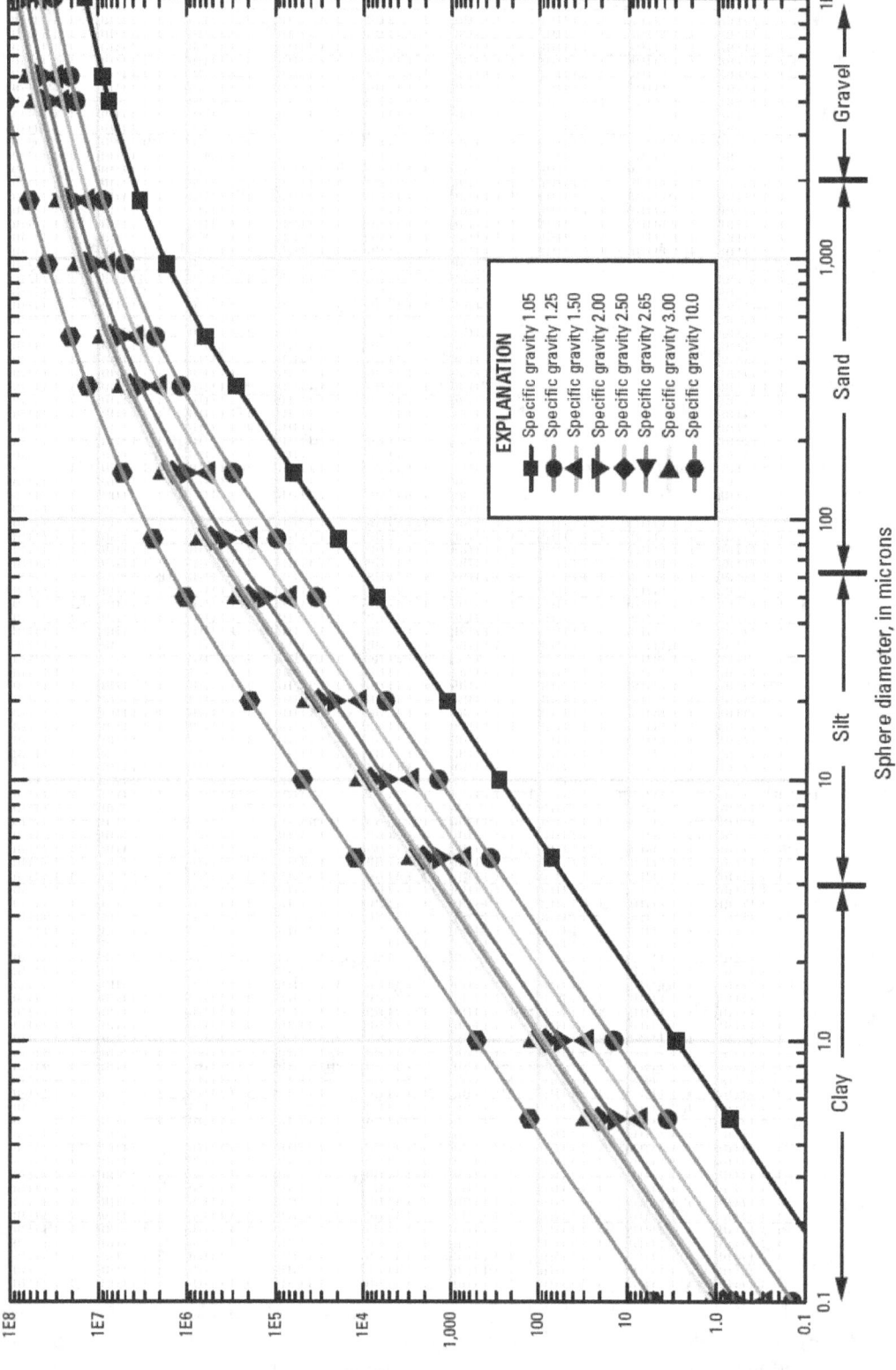

Figure 30.   The fall velocities of spheres with various specific gravities and diameters in quiescent water at 20° Celsius. Fall velocities were calculated by using the equation by Gibbs and others (1971). 1E4 is scientific notation for 10,000.

mass in the table presented by Gottschalk (1961) were about 39, 39, and 22 percent for clay, silt, and sand, respectively. Gottschalk (1961) also indicated that grain-size distributions can differ by several orders of magnitude in a reservoir with a high proportion of sand near the inlets to the reservoir and high proportions of silt and clay in other parts of the reservoir.

The settling velocities of suspended sediments are a function of the densities of the water and of the particulates (equation 30). The density of water is a function of temperature, but the change in density over the temperature range common in rivers and streams is almost negligible (Lide, 1997). Dissolved and suspended solids also may affect the density of solution, but these effects are not usually of concern unless concentrations reach 7,000 mg/L for dissolved solids (Hem, 1992) and 8,000 mg/L for suspended solids (American Society for Testing and Materials, 2009). Figure 30 includes calculated fall velocities for spheres in quiescent water with densities ranging from 1.05 to 10 g/cm^3.

Reported sediment-density values in the literature range considerably, indicating that the fall velocities of runoff sediments will range considerably. For example, the densities of many natural minerals range from 2 to 3 g/cm^3 (Perry, 1963). Sansalone and Tribouillard (1999) found that sediment densities in pavement runoff ranged from about 2.75 to 2.9 g/cm^3 over a range of grain sizes from 20 to 5,000 microns. Similarly, Li and others (2006) reported densities of sediment samples from 1.89 to 2.86 g/cm^3 and differences among grain sizes. Lin (2003) examined the density of sediment in samples of runoff from Baton Rouge and New Orleans, Louisiana, and Little Rock, Arkansas; he found consistent densities of about 2.5 g/cm^3 for urban and highway sediments in a grit chamber. Densities of sediments in a subsequent collection chamber increased from about 1.5 to 2.6 g/cm^3 with decreasing grain sizes in the range of 9,500–425 microns and then remained constant at about 2.6 g/cm^3 for grain sizes down to 25 microns. Lin (2003) noted that much of the large size fraction in runoff was composed of organic matter such as leaves and other plant materials, wood pieces, straw, and tire debris, and noted that such materials have densities of about 1.9 g/cm^3. Butler and others (1996) found that the densities of organic solids in storm sewers are in the range of 1.1 to 2.5 g/cm^3. Edil and Bosscher (1994) reported densities of tire shreds in the range of 1.02 to 1.36 g/cm^3 depending on the amount of glass belting or steel wire in the tire particles. Fowler (2008) found that sediment in parking-lot runoff samples had bulk specific gravities of 1.86 to 2.96 with a median of 2.04. Asphalt has a density of about 1.1 to 1.5 g/cm^3 (Lide, 1997). The densities of commercial-grade aluminum, bronze, copper, iron, nickel, lead, tin, and zinc are about 2.8, 8.15, 8.94, 7.86, 8.89, 11.34, 7.35 and 7.14 g/cm^3, respectively (Perry, 1963; Lide, 1997). Thus, the metal particulates in runoff will settle more rapidly than quartz particles of equivalent size and shape because the densities of commercial-grade metals are larger.

Temperature can have a substantial effect on the fall velocities of sediments, mainly because of the effect of temperature on the viscosity of water. Two factors in equation 30, the dynamic viscosity and the density of fluid, are functions of temperature. The dynamic viscosity of water at 0 °C (0.01793 poise) is about 1.37, 1.8, 2.25, and 2.7 times the viscosity of water at 10, 20, 30, and 40 °C, respectively (Lide, 1997). The density of water, however, changes by less than 1 percent over the range from 0 to 40 °C (Lide, 1997). Figure 31 indicates the effects on the fall velocity of spheres in quiescent water of changes in dynamic viscosity and water density at temperatures ranging from 0 to 30 °C. The smaller grain sizes are affected more than the larger sizes, and the curves converge toward common values for sediment in the range of coarse sands. Fang and Stefan (1999) report the minimum, maximum, and average lake temperatures at 209 locations in the conterminous United States. They provide temperatures for these lakes at 1 m below the surface and 1 m above the deepest point in the lake. Nationally, latitude and elevation are explanatory variables for the range and mean values for these temperatures. Fang and Stefan (1999) report maximum daily near-surface temperatures in the range of 19.5 to 32.8 °C, minimum daily near-surface temperatures in the range of 0 to 19.6 °C, maximum daily near-bottom temperatures in the range of 19.5 to 31.8 °C, and minimum daily near-bottom temperatures in the range of 4 to 7.7 °C.

Fall velocities of sediment and solids under natural conditions also may be affected by particle shape and by oscillations in the water column. Guy (1977) provides data on the effect of variations in particle shape and water temperature on fall velocities. Carmichael (1982) studied different particle shapes and found that settling velocities of nonspherical particles were 67 to 97 percent of the settling velocity of an equivalent sphere. The largest reductions of about 30 percent of the settling velocity were measured for disk-like particles. Reductions of 10 to 20 percent were common for cylindrical particles falling with the long axis vertical. Velocities became maximal as the aspect ratio of the particle approached 1. Similarly, Dietrich (1982) studied fall velocities of sand particles with different shape factors and found that the fall velocities of the natural sand particles used in his study were about 68 percent of the fall velocities of equivalent spheres. Relations between fall velocities and oscillations in the water column are complex, but reductions in velocities can be as much as 50 percent of the fall velocity in a quiescent fluid (Granato, 1992).

A single attenuation factor is needed for use with the SELDM lake module. A size and density-weighted attenuation factor can be calculated on the basis of the proportion of each sediment class and a representative settling velocity. If equation 30 is applied with the fall velocity of each size fraction, and the sums of the input concentrations and of the estimated lake concentrations are calculated, then an effective attenuation factor can be calculated:

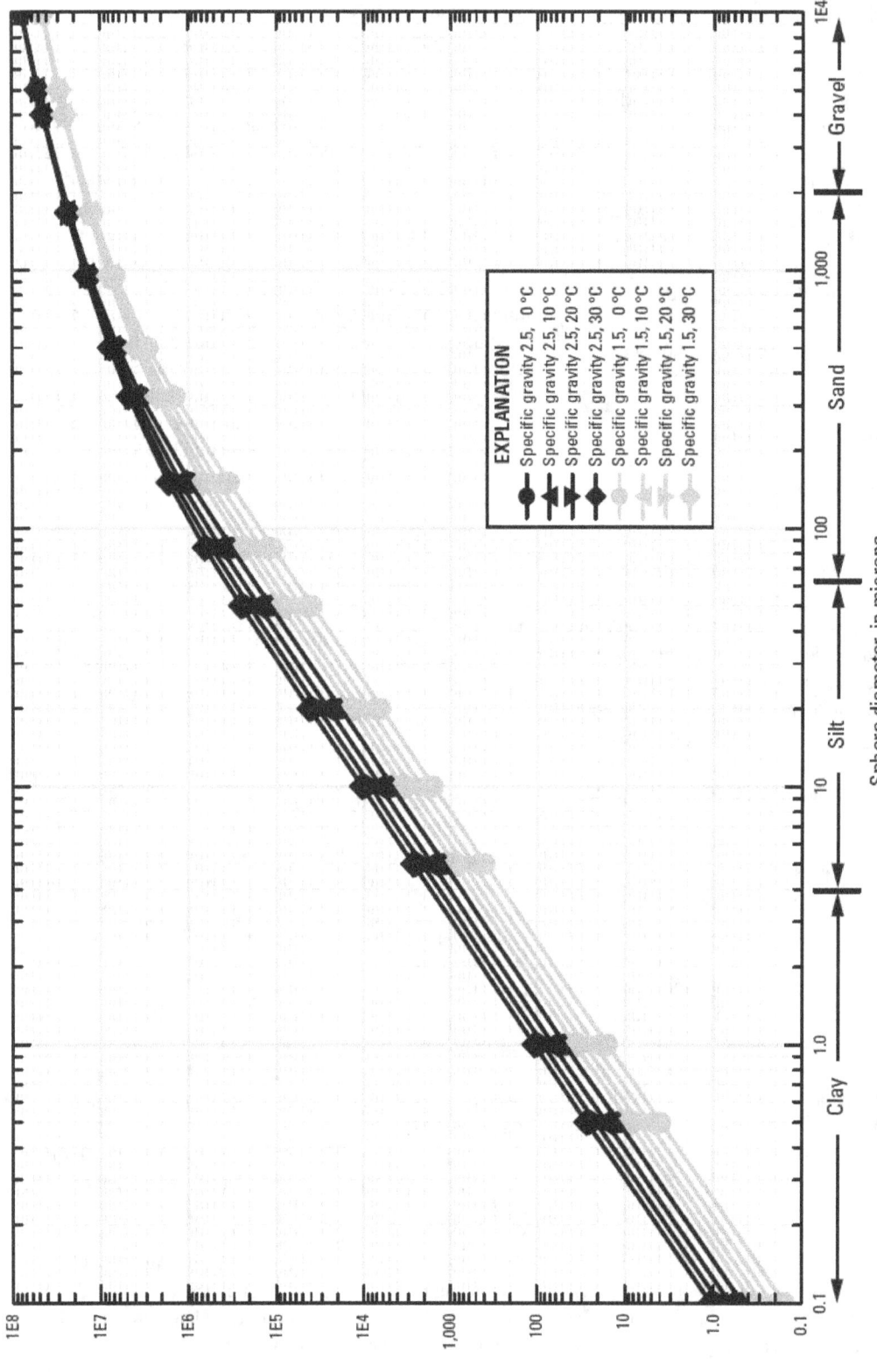

Figure 31. The fall velocities of spheres in quiescent water at 0, 10, 20, and 30° Celsius with specific gravities of 1.5 and 2.5 grams per cubic centimeter and different grain-size ranges. Fall velocities calculated by using the equation by Gibbs and others (1971). 1E4 is scientific notation for 10,000.

$$K_{Effective} = \frac{\left(\dfrac{\sum C_{in(i)}}{\sum C_{lake(i)}} - 1\right)}{T_w}, \qquad (31)$$

where

$K_{Effective}$ is the effective attenuation factor;

$C_{in(i)}$ is the ith proportion of the input concentration;

$C_{lake(i)}$ is the ith proportion of the lake concentration calculated for each fraction by using equation 28; and

T_w is the mean hydraulic residence time of water in years, calculated by using equation 24.

Table 5 shows a hypothetical example that demonstrates the method for using equation 31 to calculate a bulk sediment-attenuation factor. The spreadsheet Table5Example.xls is provided on the CD–ROM accompanying this report to facilitate analysis of attenuation factors. First, list the properties of the lake that are needed for use in equations 24, 28, and 30. Next, list the properties of the sediment that are needed for use in equation 30. In this case, we are modeling an annual average particle-size distribution. For this example, the average grain-size distribution published by Selbig and others (2011) for stormwater runoff from a mixed-use area was used as the basis for calculation of the inflow-proportion values (C_{in}). The particle diameter is selected to be at or near the minimum diameter associated with each fraction so that the calculated attenuation factors will be conservatively low. In this example, the fall velocities for spheres in quiescent water are calculated by using equation 30, and then these values are reduced by 30 percent to approximate the potential effects of nonideal conditions (non-spherical shapes and water-column oscillations). The adjusted fall velocity for each size class is multiplied by the mean depth of the lake to calculate the attenuation factor. The proportion of each size class remaining in the lake (C_{lake}) is calculated by using equation 28. In the first example shown in table 5, the sand fractions are effectively removed because of the settling velocity, depth, and hydraulic residence time of the lake. The inflow and outflow proportions are added, and the effective attenuation factor (K) is calculated by using equation 31.

This method could be applied to a less detailed grain-size distribution, and entries could be included for different particle densities within each grain-size fraction. Such an approach may or may not be warranted based on available data and the variability of particle-size distributions and particle densities in the inflows to the lake. This user-defined effective attenuation factor is used in SELDM with the stochastically generated population of annual highway-runoff and lake-basin loads to calculate a population of annual average lake concentrations.

Trace Elements

Attenuation factors for trace elements, including the highway-associated metals, are dependent on many complex and interrelated factors. Trace elements in the aquatic environment commonly are described as being in a four-phase system in association with dissolved constituents, suspended sediment, bed sediment, and biological tissue (Breault and Granato, 2003). Partitioning among these fractions depends on the physical and chemical characteristics of the water, sediments, and biota. Trace elements can move back and forth between these phases depending on the geochemistry, concentration of sediments, and properties of sediments in the solution. The total concentration of a trace element and the amount in each phase may be heavily influenced by the sample collection, processing, and analysis methods as well as the methods used to interpret available data (Breault and Granato, 2003). The results of sampling and partitioning studies of trace elements also may be affected by changes in the geochemistry of the solution between collection and processing. If the time period for the lake analysis is assumed to be sufficient to produce geochemical and biochemical equilibrium with respect to annual average concentrations, then a planning-level attenuation factor can be calculated. An attenuation factor can be calculated by estimating the proportion of the total concentration of the trace element associated with each settleable fraction of particulates.

The dissolved fraction of a trace element commonly is operationally defined as the portion associated with the water that passes through a 0.45-micron filter. This size fraction includes fine clays, colloids, and microscopic organisms; the fall velocities of such particulates, however, are negligible (figs. 30 and 31). The concentrations of dissolved trace elements at geochemical and biochemical equilibrium are approximately constant, so the attenuation factor for this portion could be modeled as equal to 0.

Highway- and urban-runoff studies have presented a wide range of the dissolved and particle-associated fractions for each of the highway-associated trace elements (Breault and Granato, 2003). For example, Sansalone and Buchberger (1997) indicated that the dissolved percentages of Cd, Cu, Ni, Pb and Zn in highway runoff were about 54–95, 31–71, 47–78, 27, and 54–96, respectively. Pitt and Vorhees (2003) indicated that the dissolved percentages of Cu, Pb and Zn in urban runoff were about 1–86, 2–20, and 1–100, respectively. Li and others (2008) summarized the results of different studies indicating that the dissolved percentages of Cd, Cu, Ni, Pb and Zn in highway runoff had ranges of 0.8–96, 1–71, 9–61, 0–45, and 4–96, respectively.

Studies of natural waters also indicate large potential differences in dissolved and particle-associated fractions, both spatially and temporally. For example, Shafer and others (1999) studied the dissolved and sediment-associated fractions of Cd, Cu, Pb, and Zn in 14 rivers in four ecoregions in Wisconsin. They determined that the average dissolved fractions were about 61, 76, 25, and 35 percent, respectively,

Table 5. Example for calculating the effective sediment-attenuation factor for a hypothetical lake.

[The example grain-size distribution is for stormwater runoff from mixed land use (adapted from Selbig and others, 2011). Concentrations are normalized by total concentration. The nonideal settling velocity adjustment is used to account for potential reductions caused by nonsphericity of sediment grains and by water-column oscillations. cm, centimeters; ft, feet; ft³/s/mi², cubic foot per second per square mile; g, grams; mi, miles; mi², square miles; r, radius; s, second]

Lake water temperature (degrees Celsius)	Water density (P_f in g/cm³)	Dynamic viscosity (η, poises)	Lake surface area (SA, in acres)	Mean lake depth eq. 25 (D, in ft)	Basin drainage area (DA, in mi²)	Long-term annual average flux of water eq. 31 (ft³/s/mi²)	Mean hydraulic residence time eq. 24 (T_w, in years)
10	0.9997	0.01307	25	16	1	1	0.5521

Sediment grain size	Initial proportion of total	Estimated particle diameter (2r, in microns)	Sediment density (P_s, in g/cm³)	Sediment settling velocity eq. 30 (V_s, in ft/year)	Nonideal settling-velocity adjustment 30.0%	Sediment-attenuation factor eq. 31 ($K_{Effective}$, in 1/year)	Remaining proportion eq. 28 (normalized C_{lake})
Medium clay	0.06	1	2	43	30	2	0.02851
Coarse clay	0.07	2	2	173	121	8	0.01292
Very fine silt	0.10	5	2	1,079	755	47	0.00371
Fine silt	0.18	8	2	2,762	1,933	121	0.00265
Medium silt	0.15	14	2	8,451	5,916	370	0.00073
Coarse silt	0.05	32	2	43,926	30,748	1,922	0.00005
Very fine sand	0.09	63	2	167,189	117,032	7,315	0.00002
Fine sand	0.08	125	2	615,975	431,183	26,949	0.00001
Medium sand	0.11	250	2	2,020,946	1,414,662	88,416	0
Coarse sand	0.11	500	2	5,363,586	3,754,510	234,657	0
Sum (normalized C_m):	1.00					Sum (normalized C_{lake}):	0.04860
Effective K eq. 31:	35						

but found order-of-magnitude differences in the speciation of trace elements from site to site. Shafer and others (1999) also include datasets from the northeastern United States and Texas with dissolved-fraction ranges of about 60–82, 10–18, and 20–49 percent for Cu, Pb, and Zn, respectively. Horowitz and others (2008) intensively monitored the water quality in 10 urban streams in Atlanta, Georgia, and found that about 72, 97, 84, 90, 98, 85, 56, and 85 percent of the annual loads of Cd, Cr, Cu, Ni, Pb, total N, total P, and Zn, respectively, were associated with suspended sediments. Therefore, about 28, 3, 16, 10, 2, 15, 44, and 15 percent of the annual loads of these elements may be operationally defined as being in the dissolved fraction.

The concentrations of sediment-associated trace elements commonly are estimated by measuring the concentrations in the sediment and multiplying these values by the concentration of suspended sediment in the water (Breault and Granato, 2003). Smith and Granato (2010) demonstrated that measured concentrations of trace elements and organic chemicals in highway sediments could be used with suspended-sediment concentrations in runoff to estimate the total concentrations in runoff. Although these estimates were biased low, the differences between calculated and measured runoff concentrations were consistent with the expected amount of the dissolved fraction. This approach could be used with the many sources of data documenting the concentrations of trace elements in natural and contaminated soils and sediments (Shacklette and Boerngen, 1984; Smith, 2006; Rice, 1999; Breault and Granato, 2003; Mahler and others, 2006; Chalmers and others, 2007). For example, Rice (1999) documented analysis of trace-element concentrations from 541 streambed-sediment samples collected from study areas across the conterminous United States and found that concentrations of trace elements in fine-grained sediments (grain sizes less than 63 microns) ranged across three to four orders of magnitude with maximum values that were 90 to 4,500 times the minimum values. The median concentrations of Cd, Cr, Cu, Pb, Ni, and Zn documented in this study were 0.4, 64, 27, 27, 27, and 110 micrograms per kilogram (µg/kg) of sediment, respectively. Rank correlations in this dataset between concentrations of these metals in bottom sediments and population densities are 0.25, 0.33, 0.49, 0.61, 0.32, and 0.49 for the same six elements, respectively. Chalmers and others (2007) then used these data to develop relationships between land-use characteristics and sediment quality.

Information about the masses of trace elements associated with different grain-size fractions is useful for calculating trace-element attenuation factors because of the large differences in settling velocities of different fractions. It is generally accepted that the fine fractions of sediments are associated with higher concentrations of trace elements than are the larger grain-size fractions. Horowitz and Elrick (1987) measured metal concentrations and the properties of sediment in the grain-size intervals 0 to less than 2, 2 to less than 16, 16 to less than 63, and 63 to less than 125 microns in samples collected from rivers and lakes across the United States.

They found that correlations with metal concentrations were strongest in the fractions representing 16 to less than 63 and 63 to less than 125 microns and attributed this to the organic fraction, surface area, and surface chemistry of these grain-size fractions. Data from Li and others (2006) indicate that concentrations of copper, lead, and zinc in the fine fractions (clay and silt) of urban runoff were about four times the concentrations in the sand fraction, but because of the grain-size distribution, about 20, 30, and 50 percent of the metals (by mass) could be associated with the clay, silt, and sand fractions, respectively.

The data in table 6 provides a hypothetical example that demonstrates the method for calculating a bulk attenuation factor for a total trace-element concentration, which is defined as including dissolved and particle-associated fractions. The spreadsheet Table6Example.xls is provided on the CD–ROM accompanying this report to facilitate analysis of attenuation factors. First, an effective sediment attenuation factor is calculated by using equation 31. This example is similar to the first example in table 5, but only four grain-size fractions were calculated for simplicity. The dissolved fraction is assumed to be conservative and therefore has an attenuation factor of 0. The proportion of the trace element in each particle-size fraction remaining in the lake (C_{lake}) is calculated by using equation 28. The proportions of the total trace element associated with the sand fractions in this example are effectively removed because of the settling velocity, depth, and hydraulic residence time of the lake. The inflow and outflow proportions are added, and the effective attenuation factor (K_{eff}) is calculated by adding the dissolved portion to the remaining sediment fractions and using equation 31. The user-defined effective attenuation factor for the trace element can be used in SELDM with the stochastically generated population of annual highway and lake-basin loads to calculate a population of annual average lake concentrations.

Organic Chemicals

Attenuation factors for organic chemicals depend on many factors including particulate settling, volatilization, biodegradation, photolysis, and hydrolysis (U.S. Environmental Protection Agency, 1985a, b, c, 1986a, 1987; Thomann and Mueller, 1987). Lopes and Dionne (2003) provide a review of the occurrence and distribution of organic chemicals that are of primary concern for many highway and urban-runoff studies. Smith and Granato (2010) provide data on the concentrations of organic chemicals in highway runoff, suspended sediment, and plant matter from highways in Massachusetts. Many organic chemicals of concern in urban and highway runoff are associated primarily with suspended particulates. Several regional and national studies provide information on the concentrations of many organic chemicals in stream and lake-bed sediments in areas of different land use throughout the United States with explanatory variables for estimating sediment concentrations (Yorke and others, 1985; Van Metre and others, 2000; Van Metre and Mahler,

Table 6. Example for calculating the effective sediment- and trace-metal-attenuation factors for a hypothetical lake.

[The example grain-size distribution is for stormwater runoff from mixed land use (adapted and simplified from Selbig and others, 2011). Concentrations are normalized by total concentration. The nonideal settling velocity adjustment is used to account for potential reductions caused by nonsphericity of sediment grains and by water-column oscillations. cm, centimeters; ft, feet; ft³/s/mi², cubic foot per second per square mile; g, grams; mi, miles; mi², square miles; r, radius; s, second]

Lake water temperature (degrees Celsius)	Water density (P_f, in g/cm³)	Dynamic viscosity (η, poises)	Lake surface area (SA, in acres)	Mean lake depth eq. 25 (D, in ft)	Basin drainage area (DA, in mi²)	Long-term annual average flux of water (ft³/s/mi²)	Mean hydraulic residence time eq. 24 (T_w, in years)
10	0.9997	0.01307	25	16	1	1	0.5521

Sediment grain size	Initial proportion of total	Estimated particle diameter (2r, in microns)	Sediment density (P_s, in g/cm³)	Sediment settling velocity eq. 30 (V_s, in ft/year)	Nonideal settling-velocity adjustment 30.0%	Sediment-attenuation factor eq. 31 ($K_{Effective}$, in 1/year)	Remaining proportion eq. 28 (normalized C_{lake})
Clay	0.26	1	2	43	30	2	0.12356
Silt	0.60	8	2	2,762	1,933	121	0.00885
Fine sand	0.08	63	2	167,189	117,032	7,315	0.00002
Medium sand	0.06	250	3	2,772,380	1,940,666	121,292	0
Sum (normalized C_{in}):	1.00					Sum (normalized C_{lake}):	0.13243

Effective sediment K eq. 31:   12

Trace metal fraction	Initial proportion of total	Sedimentation attenuation factor K (1/year)	Remaining proportion equation 28
Dissolved	0.2	0	0.20000
Clay	0.4	2	0.19009
Silt	0.2	121	0.00295
Fine sand	0.15	7,315	0.00004
Medium sand	0.05	121,292	0.00000
Sum (normalized C_{in}):	1.00		0.39308

Sum (normalized C_{lake}):

Effective trace element K eq. 31:   3

2005; Chalmers and others, 2007). A large proportion of the total attenuation factor (equation 29) for these organic compounds may be associated with sedimentation. The U.S. Environmental Protection Agency (1987), however, provides guidance and values for estimating general first-order decay rates, as well as the solubility, potential volatilization, biodegradation, photolysis, and hydrolysis of many organic chemicals for calculating a total attenuation factor.

Interpreting the Results of an Analysis

Water-resource managers are concerned about the frequency, magnitude, and duration of concentrations and loads that may have an adverse effect on the quality of receiving waters (Driscoll and others, 1979, 1989; Athayde and others, 1983; Di Toro, 1984; Driscoll, Shelley, and others, 1989; U.S. Environmental Protection Agency, 1996b, 2002b, 2007a; Smith and others, 2001; Borsuk and others, 2002; Bonta and Cleland, 2003; Gibbons, 2003; Novotny, 2004; Elshorbagy and others, 2007; Brouwer and De Blois, 2008; Langseth and Brown, 2011). There has been a growing awareness that statistical approaches and Monte Carlo methods are needed to address these concerns. SELDM generates a long record of storm-event and annual concentrations, flows, and loads that match input statistics so that scientists, engineers, and decisionmakers can assess the potential risks of water-quality excursions and the potential effects of mitigation measures to reduce those risks. SELDM also calculates the relative contribution of the highway runoff to such excursions. The Monte Carlo methods used by SELDM to generate storm-event statistics are necessary for quantitative analysis of risk because simple methods are not sufficient to characterize the interplay of different distributions for precipitation, prestorm flow, runoff coefficients, concentrations, and BMP-performance measures. Simpler methods may provide estimates of mean values, but it is commonly the extreme events that are, or should be, of most interest to scientists, engineers, and decisionmakers.

SELDM produces one documentation file, eight stream-package output files, and one lake-package output file that document the inputs to each analysis and provide the results of analysis. The numerical outputs are in a tab-delimited text format that is readily copied from the output files and into spreadsheet programs, graphing packages, statistical software, and word-processing tables. This output is designed to facilitate post-modeling analysis and presentation of results. Details about the format and content of each file are described in the section of this manual on model-output files.

The benefit of the Monte Carlo analysis is not to decrease uncertainty in the input statistics, but to represent the different combinations of the variables that determine potential risks of water-quality excursions. Uncertainty in input statistics can be expressed as confidence limits for sample statistics, correlation coefficients, and regression relations used to estimate such statistics (Haan, 1977; Chow and others, 1988; Press and

others, 1992; Helsel and Hirsch, 2002). Statistical confidence limits most commonly are a function of 1 over the square root of the number of samples. For example, the standard error of the skew coefficient is equal to the square root of the quotient of six divided by the number of samples (Press and others, 1992). If data are lognormal, the real coefficient of skew would be equal to 0, but the 95-percent confidence intervals for this statistic would be plus or minus 2.15, 1.52, 1.24, and 1.1 if the sample sizes were 5, 10, 15, and 20, respectively. SELDM is designed to generate about 800 to 2,400 runoff-producing storm events based on the average number of storm events per year to stabilize the statistics of output results (appendix 1).

The 95-percent confidence limits of the SELDM outputs, however, are not more precise than the inputs because the outputs are based solely on inputs. For example, the average, standard deviation, and skew of the logarithms of total phosphorus data measured in 18 highway-runoff samples collected at USGS station 423027071291301 along State Route 2 in Littleton, Massachusetts, were -1.05, 0.423, and -0.679, respectively (Smith and Granato, 2010). SELDM results based on these statistics indicate that 0.5 percent of storms would have concentrations exceeding 0.6 mg/L. However, the theoretical 0.5-percent exceedances calculated by the frequency-factor method with the Pearson Type III variates would be 0.42, 1.1, and 3.02 for skews of -1.13 (the lower 95-percent confidence limit for a sample of 18 values from a lognormal distribution), 0 (the expected skew of a lognormal distribution), and 1.13 (the upper 95-percent confidence limit for a sample of 18 values from a lognormal distribution), respectively.

SELDM does, however, provide a method for rapid assessment of information that is otherwise difficult or impossible to obtain because it models the interactions among hydrologic variables (with different probability distributions) that result in a population of values that represent likely long-term outcomes from runoff processes and the potential effects of different mitigation measures. SELDM also provides the means for rapidly doing sensitivity analyses to determine the potential effects of different input assumptions on the risks for water-quality excursions.

SELDM produces a population of storm-event and annual values to address the questions about the potential frequency, magnitude, and duration of water-quality excursions. The output represents a collection of random events rather than a time series. Each storm that is generated in SELDM is identified by sequence number and annual-load accounting year. The model generates each storm randomly; there is no serial correlation, and the order of storms does not reflect seasonal patterns. The annual-load accounting years, which are just random collections of events generated with the sum of storm interevent times less than or equal to a year, are used to generate annual highway flows and loads for TMDL analysis and the lake basin analysis.

Probability plots (sometimes called duration curves, quantile plots, or percentile plots) provide one of the simpler methods for evaluating and presenting the results of statistical analyses (Riggs, 1968; Haan, 1977; Chow and others, 1988; Helsel and Hirsch, 2002; U.S. Environmental Protection Agency, 2007a). A probability plot is a scatterplot in which the magnitudes of data values are graphed with respect to a statistical estimate of the probability of occurrence. The statistical estimate of the probability of occurrence may be expressed as an exceedance probability, a cumulative probability, a normal variate, or a return period.

An exceedance probability indicates the proportion or percentage of values that equal or exceed a given value. Data are sorted in descending order to graph the values with respect to the exceedance probabilities. A cumulative probability indicates the proportion or percentage of values that are less than or equal to a given value. Data are sorted in ascending order to graph the values with respect to the cumulative probabilities. Exceedance probabilities are equal to 1 minus the associated cumulative probability (100 minus the associated cumulative probability if percentiles are used).

The normal variates indicate the distance, measured as the number of standard deviations, of each percentile from the mean under the assumption that the median equals the mean in a standard normal distribution. Return periods are the reciprocals of the probability of occurrence. Return periods commonly are expressed as a number of years; however, a return period is not a literal expectation of the time between values of the given magnitude. For example, a three-year concentration value has a 33-percent chance of occurring in any given year but may be equaled or exceeded many times in any given year.

Probability plots may be constructed by using a linear axis or a probability axis. A probability axis can be constructed for a particular probability distribution so that the probability plot of data that fit the selected distribution will fall on a straight line (Haan, 1977); however, the normal probability distribution is the one most commonly implemented in graphing software. If data are lognormally distributed, then the plotted logarithms (or untransformed data plotted on a logarithmic scale) will form a straight line. Probability axes facilitate examination of the extreme percentiles that are of interest for hydrologic applications.

The example problem that was included in the SELDM database is used to demonstrate inputs to and outputs from the model. This example is based on the highway-runoff-monitoring site described by Gupta and others (1981) on a section of I–81 near Harrisburg, Pennsylvania, because the highway site and the upstream basin are well documented in that report. The site is in ecoregion 67 (Central Appalachian Ridges and Valleys) and the mid-Atlantic rain zone (no. 3). The area of the highway site is 18 acres, the impervious fraction of the area is 0.27, and the drainage length is about 2,000 ft. The area of the upstream basin is about 0.5 mi², the impervious fraction of this area is 0.007, and the basin length is about 6,500 ft.

Total phosphorus EMCs (USEPA parameter code p00665) were selected to demonstrate the analysis and presentation of storm-event outputs and annual outputs from SELDM. Total phosphorus EMCs measured in 18 highway-runoff samples collected at USGS station 423027071291301 along State Route 2 in Littleton, Mass. (Smith and Granato, 2010), were used to represent recent (2005–07) runoff quality at the I–81 site because the Route 2 and I–81 sites have similar highway characteristics. Upstream concentrations at the test site were modeled by using the planning-level transport curve developed by Granato and others (2009) with data from this ecoregion.

Storm-Event Results

This storm-event example is used to demonstrate general concepts for developing and interpreting the probability plots that can be used to interpret results of a SELDM analysis. This example focuses on interpretation of concentration outputs from SELDM; similar methods would be employed to interpret results for storm-event characteristics, prestorm flows, runoff coefficients, stormflows, and loads. Figure 32 demonstrates construction of probability plots for the highway-runoff EMC outputs with three different probability axes. Concentration data are plotted on a logarithmic axis because the population can be characterized by a log-Pearson type III distribution based on the mean, standard deviation, and skew of the logarithms of the highway-runoff samples. Figure 32A was constructed by using a linear probability axis because commonly used spreadsheet programs do not provide the option to use a probability axis. Statistical and graphing programs, however, do commonly provide this option. A linear axis can be used to view most data in the simulated population and the range of data, but details of extreme values outside the 2nd and 98th percentiles can be difficult to distinguish. This is because probability plots of exponential, lognormal, and log-Pearson type III distributions are S-curves plotted on a logarithmic data axis and a linear probability axis and J-curves plotted on an arithmetic data axis and a linear probability axis. Extreme probability values, which commonly are of greatest interest in hydrologic studies, are difficult to distinguish because of the short range of extreme values at the ends of the linear probability axis.

Figure 32B was constructed by using a probability axis with percentile values that have been converted to standard normal variates. This method is used to create a probability axis in commonly used spreadsheet programs that do not provide the option to use a probability axis. The simulated concentration data in figure 32B are linearized by this transformation. If the simulated data were from a lognormal distribution, they would plot on a straight line. The logarithms of the concentration data have a slight negative skew, so the simulated data plot has a slight curve that is concave downward (Haan, 1977; Chow and others, 1988; Granato and others, 2010). The Microsoft Excel® spreadsheet software has a function named "NORMSINV" that will convert the percentiles (scaled from 0 to 1) to the associated standard normal variates. If this function is not available, the

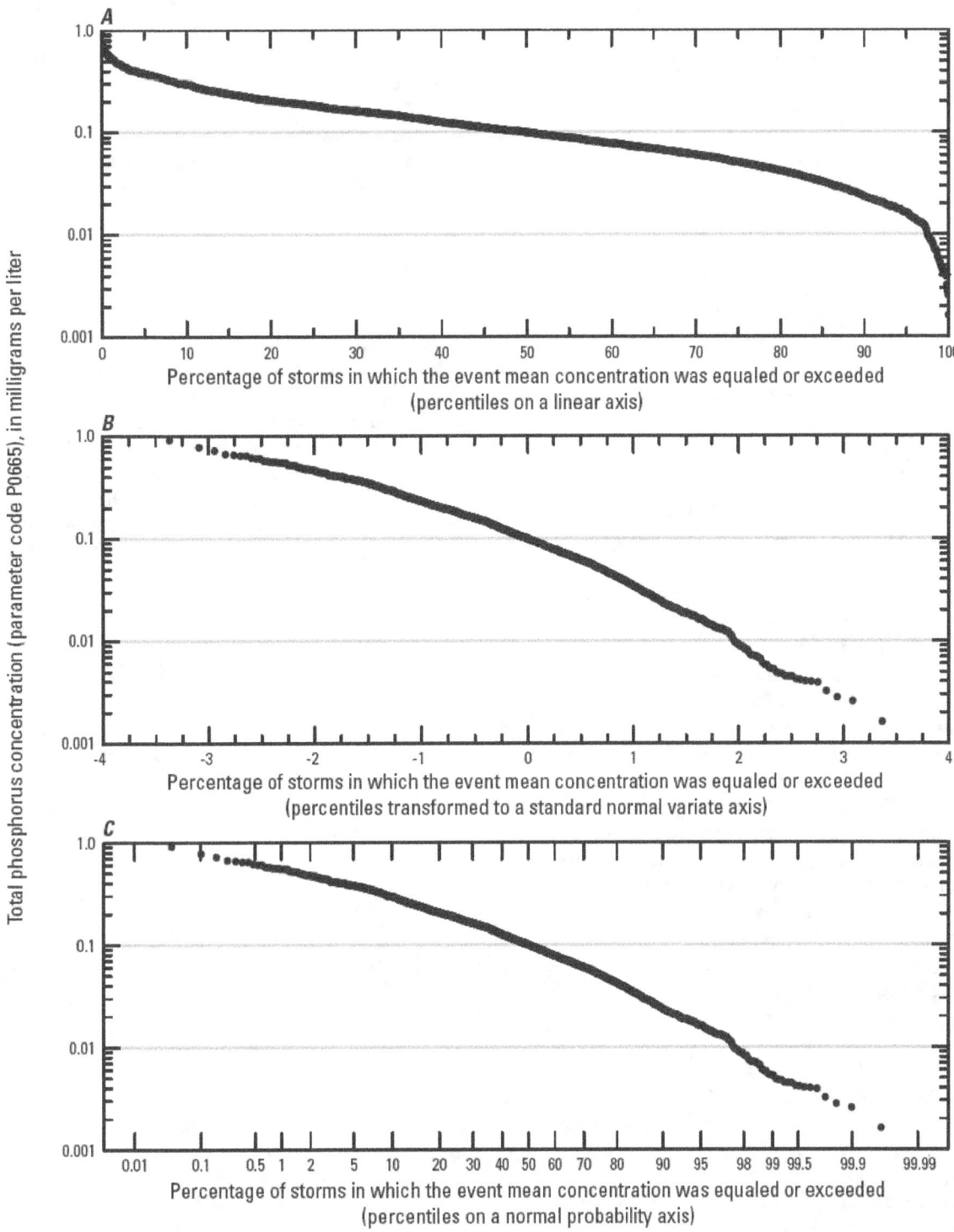

Figure 32. The stochastic population of total phosphorus concentrations calculated with statistics from highway-runoff data collected on State Route 2 in Littleton, Massachusetts (Smith and Granato, 2010). The graphs show concentrations plotted by *A*, percentiles on a linear axis, *B*, percentiles transformed to the associated standard normal variates, and *C*, percentiles on a normal probability axis.

conversion can be done by a two-step algebraic approximation (Abramowitz and Stegun, 1965; Chow and others, 1988). First, an intermediate variable w is calculated from the probability p (scaled from 0 to 1) as

$$w = \left(\ln\left(\frac{1}{p^2} \right) \right)^{0.5}$$ (32)

if p is less than or equal to 0.5 and

$$w = \left(\ln\left(\frac{1}{(1-p)^2} \right) \right)^{0.5}$$ (33)

if p is greater than or equal to 0.5. Second, w is used to calculate the normal variate Z with the algebraic approximation

$$Z = w - \frac{2.515517 + 0.802853w + 0.010328w^2}{1 + 1.432788w + 0.189269w^2 + 0.001308w^3} .$$ (34)

If p is greater than 0.5, then Z is given a negative sign; the maximum error in this approximation is 0.00045 (Abramowitz and Stegun, 1965; Chow and others, 1988).

The standard normal variate axis can be converted to a probability axis manually by methods described by Haan (1977). The data in figure 32C are plotted on a probability axis with percentiles spaced according to the normal distribution. Comparison with figure 32B shows that the spacing of percentile values is the same. Whereas both plots clearly show the extreme values between 0 and 2 percent and between 98 and 100 percent, the total extent of these ranges along the X axis—about 50 percent of the entire length—may give the appearance of a greater number of extreme points to a casual observer.

The EMCs are plotted in descending order, so the graphs indicate the percentage of storms in which any given concentration was equaled or exceeded, whereas most hydrologic probability plots indicate the percentage of time a given concentration, flow or load was equaled or exceeded. This is an important distinction if SELDM outputs are to be evaluated with respect to water-quality criteria that are based on the frequency and duration of water-quality excursions. Evaluation of the SELDM outputs indicates that, for one Monte Carlo run, the runoff-producing storm-event durations for 1,586 storms represented about 4.67 percent of the total time during a 30-year period, and highway runoff flowed for about 5.29 percent of the time (these percentages would be expected to vary from run to run at one site and from site to site). Thus, if highway-runoff quality exceeds a water-quality target during 10 percent of storms, this exceedance might occupy only about 0.5 percent of the time during a 30-year period. This analysis reveals a paradoxical effect for implementation of BMPs: if a BMP cannot substantially

reduce the concentration of a constituent but does extend the runoff hydrograph to attenuate the adverse effects of peak flows, then the BMP may contribute to a higher percentage of time-based excursions.

Figure 33 demonstrates construction of a probability plot for the same highway-runoff EMC outputs with a return-period axis (sometimes called a recurrence-interval axis). The return period is defined as the reciprocal of the probability of occurrence, commonly in units of years. The longer the return period, the lower the probability of recurrence. Return-period values, however, must be interpreted with care. For example, the 100-year flood is a commonly used statistic for hydrologic design and planning. The 100-year flood, however, is not the flood that occurs only once in 100 years; it is the magnitude of maximum annual streamflow that, statistically, is expected to have a 1-percent chance of occurring in any given year. SELDM randomly produces the times between storm-event midpoints and groups storms into annual-load accounting years (defined as total times between storm-event midpoints equal to or greater than 1 year, equal to 8,760 hours for normal years and 8,784 hours for leap years). By chance, all the extreme events generated by SELDM in a Monte

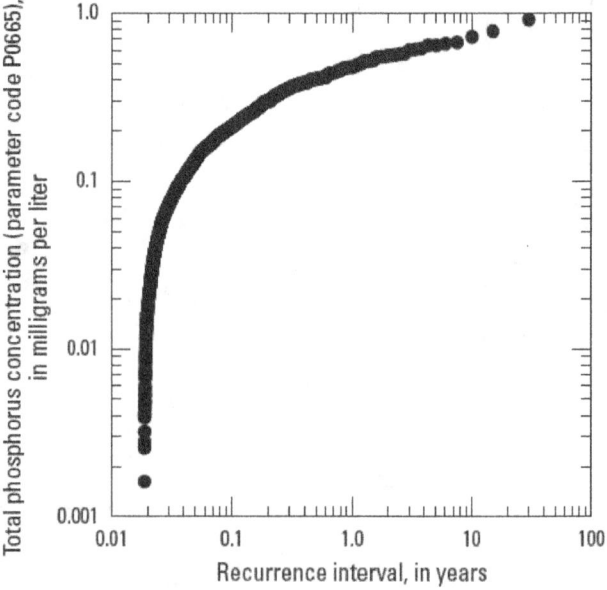

Figure 33. The stochastic population of total phosphorus concentrations calculated with statistics from highway-runoff data collected on State Route 2 in Littleton, Massachusetts (Smith and Granato, 2010). The graph shows concentrations in runoff from 1,586 simulated storm events over 30 years plotted by recurrence intervals on a logarithmic axis.

Carlo run could occur within one annual load year. This is an unlikely but plausible occurrence in real precipitation records and in Monte Carlo simulations, given a long enough period of record. Return periods commonly are calculated by using the Weibull plotting position formula (Haan, 1977; Chow and others, 1988; Helsel and Hirsch, 2002). If the SELDM output were time based rather than storm based, then the recurrence interval could be calculated as $1/(N+1)$, where N is the number of days or years in the record. SELDM, however, is storm-based, and the percentiles are calculated by using the Cunnane plotting position formula because it results in more unbiased values for normal percentiles (Helsel and Hirsch, 2002). The recurrence intervals are calculated from the Cunnane percentiles as

$$Y_R = \frac{1}{A_y \left(\dfrac{P_c(N+0.2)+0.4}{N+1} \right)} \tag{35}$$

if the data are sorted in descending order and

$$Y_R = \frac{1}{A_y \left(1 - \dfrac{P_c(N+0.2)+0.4}{N+1} \right)} \tag{36}$$

if the data are sorted in ascending order, where

Y_R is the return period for storms, in years;
P_c is the Cunnane plotting position percentile (between 0 and 1);
N is the number of storms generated in a given run of the Monte Carlo model; and
A_y is the average number of annual-load accounting years in the generated record (equal to the number of storms N divided by the number of annual-load accounting years).

Two equations are necessary because return-period plots for storm-event concentrations, flows, and loads should be constructed so that higher values have longer return periods. The concentration data in figure 33 are plotted on a logarithmic axis because of the selected probability distribution for total phosphorus concentrations. The return periods are plotted on a logarithmic axis to expand the data over the measurement range. The maximum return period for this example is about 30 years, and about 98.1 percent of the storms (the bulk of the stochastic data) have return periods that are less than 1 year. In this example, on average, the total phosphorus EMC that is expected to be equaled or exceeded once every 3 years is about 0.6 mg/L.

SELDM does not explicitly provide comparisons to water-quality target concentrations because standards may differ from state to state or represent different interpretations of national standards, may depend on water use and water

quality, and may change with time. SELDM does, however, provide the information necessary to evaluate results quantitatively. For example, if a standard for total phosphorus in stormwater discharge was 0.5 mg/L, then the output shown in the probability plot indicates that this standard would be exceeded by only about 1.6 percent of storm EMCs over a long period of time (fig. 32C), equivalent to a return-period estimate of about 1.2 year (fig. 33). If, however, the standard for stormwater discharge was set at 0.1 or 0.01 mg/L, then it would be expected that EMCs may exceed these values in about 50 percent or 97.5 percent of storms (fig. 32C). Return-period values for these percentiles are about 0.04 and 0.02 years (fig. 33).

SELDM also is designed to be a tool for evaluating the potential effectiveness of mitigation measures for reducing water-quality excursions in highway-runoff discharges and in receiving waters. SELDM can be run to evaluate the quality and quantity of discharge to the receiving water with and without use of a BMP. Figure 34 is a probability plot showing total phosphorus EMCs in highway runoff and in the effluent from a hypothetical BMP. The EMCs in BMP effluent were calculated from the highway-runoff EMCs by using the trapezoidal-distribution statistics for total phosphorus reductions in the Lake Ridge detention pond (fig. 25A). In this example, the estimated minimum irreducible concentration of total phosphorus was estimated to be 0.005 mg/L.

In general, the duration curve of the BMP-effluent concentrations is shifted downward from the curve for the highway-runoff discharges, indicating the effectiveness of the BMP for reducing concentrations. The treatment statistics for the Lake Ridge detention pond (fig. 25) result in a decrease in the maximum concentration from about 0.92 to 0.23 mg/L (fig. 34). These treatment statistics also indicate that a substantial reduction in water-quality excursions could be achieved for high allowable effluent-concentration limits, but only modest improvements could be made for low concentration limits. For example, if the effluent-concentration limit was 0.1 mg/L, then the percentage of storms exceeding this limit for untreated highway runoff would be about 50 percent, but the percentage of storms exceeding this limit for BMP effluent would only be about 3.8 percent, which is a major improvement. If the effluent limit were 0.01 mg/L, however, the percentage of storms exceeding this limit for untreated highway runoff would be about 97.5 percent, but the percentage of storms exceeding this limit for BMP effluent would only be about 95.4 percent, which is an almost imperceptible improvement. The effect of the minimum irreducible concentration of 0.005 mg/L is apparent because the BMP-effluent probability plot plateaus at this value, even if highway-runoff concentrations are lower. Thus, if the effluent concentration limit was 0.004 mg/L, this BMP could not achieve the effluent limit in this example.

A problem with probability plots is that it is easy to assume that individual storms correspond to one another on the probability axis. This would be the case if a constant factor were applied to all values. If the average effectiveness

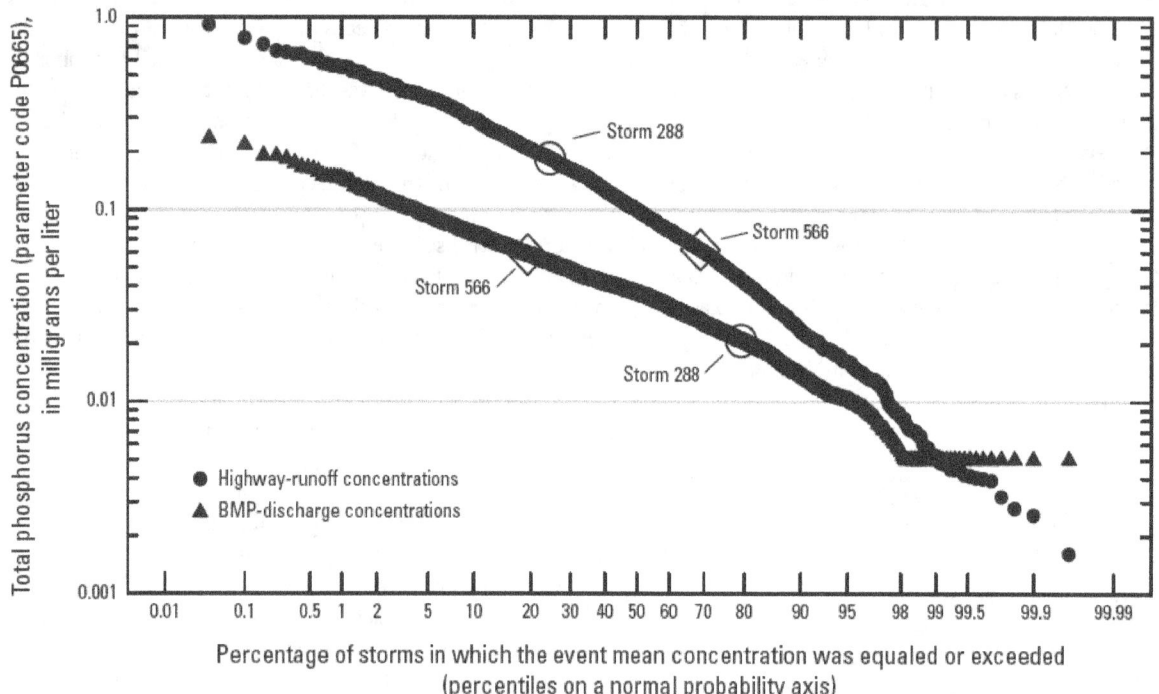

Figure 34.   The stochastic population of total phosphorus concentrations in highway runoff and in discharge from a best management practice (BMP). The highway-runoff values were calculated with statistics from data collected on State Route 2 in Littleton, Massachusetts (Smith and Granato, 2010). The BMP-discharge values were calculated by using the trapezoidal distribution statistics for the Lake Ridge detention pond (fig. 25) and a minimum irreducible concentration of 0.005 milligrams per liter.

of the BMP were used, then the BMP-effluent-concentration curve would be parallel to the runoff-concentration curve on a logarithmic scale because the outlet concentration is a constant proportion of the inlet concentration. Use of the stochastic treatment variable, however, has the effect of shuffling individual storms along the probability axis. The probability plots in figure 34, however, reflect patterns in measured BMP-performance data. The lines diverge at low percentiles because the negative correlation between the ratio of outflow to inflow concentrations and runoff concentrations indicates better treatment at higher concentrations.

The storms identified in figure 34 demonstrate the effects of the stochastic treatment ratios. These storms represent the maximum differences in plotting-position percentiles between the stochastic highway-runoff quality population and the stochastic BMP-effluent quality population. Storm 566 in this example output had a runoff concentration of about 0.062 mg/L (with a plotting-position percentile of 69 percent) and an effluent ratio of about 0.94, so the BMP-effluent concentration for this storm was 0.058 mg/L (with a plotting-position percentile of 19.4 percent). The statistics for storm

288 include a low effluent ratio of (about 0.11) with a runoff concentration of about 0.18 mg/L (with a plotting-position percentile of 24.7 percent) and a BMP-effluent concentration of 0.021 mg/L (with a plotting-position percentile of 79.3 percent).

If receiving-water concentrations are the primary concern rather than effluent concentrations, then the quality of upstream stormflows and highway (or BMP) effluent are of interest for evaluating potential mitigation measures. To estimate the quality of upstream stormflows, Granato and others (2009) developed a two-segment water-quality transport curve relating total phosphorus concentrations to streamflows for ecoregion 67. They found that total phosphorus concentrations varied randomly with a median value of 0.05 mg/L for streamflows less than 1.5 ft^3/s/mi^2 and increased substantially with increasing flow above this threshold. This relation indicates that total phosphorus in the streams in ecoregion 67 was mobilized by stormflows (Granato and others, 2009). Use of this transport curve to estimate upstream EMC values resulted in a population of concentrations that ranged from 0.007 to 2.0 mg/L, with a median of 0.104

and an average of 0.156 mg/L (fig. 35A). These upstream EMCs exceeded concentrations for both highway runoff and BMP effluent. Highway-runoff EMCs ranged from 0.002 to 0.92 mg/L, with a median of 0.1 and an average of 0.13 mg/L (fig. 35B). BMP-effluent EMCs ranged from 0.005 to 0.23 mg/L, with a median of 0.037 and an average of 0.041 mg/L (fig. 35B). Although the EMCs for the upstream stormflows were generally higher than those for highway runoff and BMP effluent in this example, the upstream EMCs for an individual storm may be higher, lower, or equal to the EMCs for highway runoff or BMP effluent.

The downstream EMCs are a function of the concentration and flow of the upstream-stormflow and highway-runoff or BMP effluent (fig. 1). The simulated dilution factors, which are defined as the highway (or BMP) discharge volumes divided by the concurrent downstream stormflows, indicate that runoff from the I–81 site commonly was a small portion of the downstream flow for most storms in this example. Dilution factors ranged from 0.02 to 61.2 percent, with a median of 7.6 percent and an average of 9.9 percent. Downstream concentrations ranged from 0.008 to 1.97 mg/L with a median of 0.11 mg/L and an average of 0.16 mg/L if highway runoff was discharged directly to the stream (fig. 35C). Downstream concentrations ranged from 0.007 to 1.97 mg/L with a median of 0.10 mg/L and an average of 0.15 mg/L if the BMP effluent was discharged to the stream instead. In this example, the BMP was highly effective for reducing high concentrations of total phosphorus in highway runoff (fig. 34), but had an almost negligible effect on the downstream EMCs for most storms. BMP reductions had a small effect on receiving-water concentrations for most storms at this site because the background EMCs were large and the highway-runoff volumes small in comparison to upstream flows.

In this example, use of the BMP at the I–81 site could have a substantial effect on downstream concentrations after some storms (fig. 35C). The results for storm 15 were atypical in comparison to most storms but indicate how random combinations of input variables can have a large effect on the outcome. Upstream stormflows for storm 15 were low (equaled or exceeded 99.5 percent of the time), and therefore the upstream EMC, which was calculated by using the water-quality transport curve, also was low (about 0.007 mg/L, which was equaled or exceeded 99.7 percent of the time) (fig. 35A). By chance, the highway-runoff volume was higher than the median (runoff volumes in about 43 percent of storms equaled or exceeded the runoff during storm 15). The difference between the EMC values for the highway runoff and BMP effluent also was large for this storm (fig. 35B). The combination of low upstream stormflow and EMC, substantial highway stormflows, and the large differences among the highway EMC, the BMP EMC, and the upstream EMC resulted in a substantial difference between the downstream EMCs with and without BMP treatment for storm 15 (fig. 35C).

Overall, however, the probability plots in figure 35C indicate that implementation of this BMP mitigation method will not make a substantial difference in water-quality excursions of total phosphorus at this site. For example, upstream EMCs equaled or exceeded 0.01, 0.1 or 1.0 mg/L during about 99.2, 51.6, and 0.349 percent of storm events, respectively, (fig. 35A). In downstream runoff, the EMCs that resulted from untreated highway-runoff discharge equaled or exceeded 0.01, 0.1 or 1.0 mg/L during about 99.8, 53.9, and 0.343 percent of storm events, respectively (fig. 35C). The EMCs that resulted from treated BMP discharge equaled or exceeded 0.01, 0.1 or 1.0 mg/L during about 99.5, 48.7, and 0.339 percent of storm events, respectively. In this example, the highway BMP substantially reduced concentrations in discharges to the stream, but upstream controls would be needed to make substantial changes in stream-water quality. The changes in downstream quality caused by BMP treatment are so minor that they could be altered by changing the Monte Carlo seed values. Changing the seed values would shuffle the random combinations of flows, concentrations, and treatment efficiencies among the simulated storm events. In this example, changes in downstream water quality could be altered randomly and systematically by doing a sensitivity analysis with different input statistics.

Annual Results

SELDM produces two files with annual results, the annual highway-runoff-loads output file and the lake-basin-analysis output file. The annual highway-runoff-loads output file is designed to facilitate analysis of TMDLs for the highway site and to record the annual highway contributions to the lake-basin analysis. The population of annual runoff loads from the highway indicates the potential highway contribution and the uncertainty of such estimates. The population of annual runoff loads from the BMP discharge indicates the potential for reducing the highway loads to meet any proposed load allocations. The lake-basin-analysis output file is designed to document annual inputs from the highway (with or without a BMP) and from the rest of the lake basin. SELDM uses the random samples of highway discharges and loads with random samples of discharges and loads from the entire lake basin to generate a random sample of total annual loads from the entire lake basin and average annual concentrations in the lake. The lake-basin output is based on the highway-runoff results for storm events but calculates daily loads from the rest of the lake basin for each day during the entire simulation period because a substantial proportion of many constituents can be transported during dry-weather base flows.

SELDM does not calculate annual loads or produce annual output files for the upstream stormflow or the downstream receiving water in stream-basin analyses because the storm-event analyses do not include dry periods between storms. Also, instead of recording the flows and loads for the entire upstream-runoff hydrograph, SELDM calculates

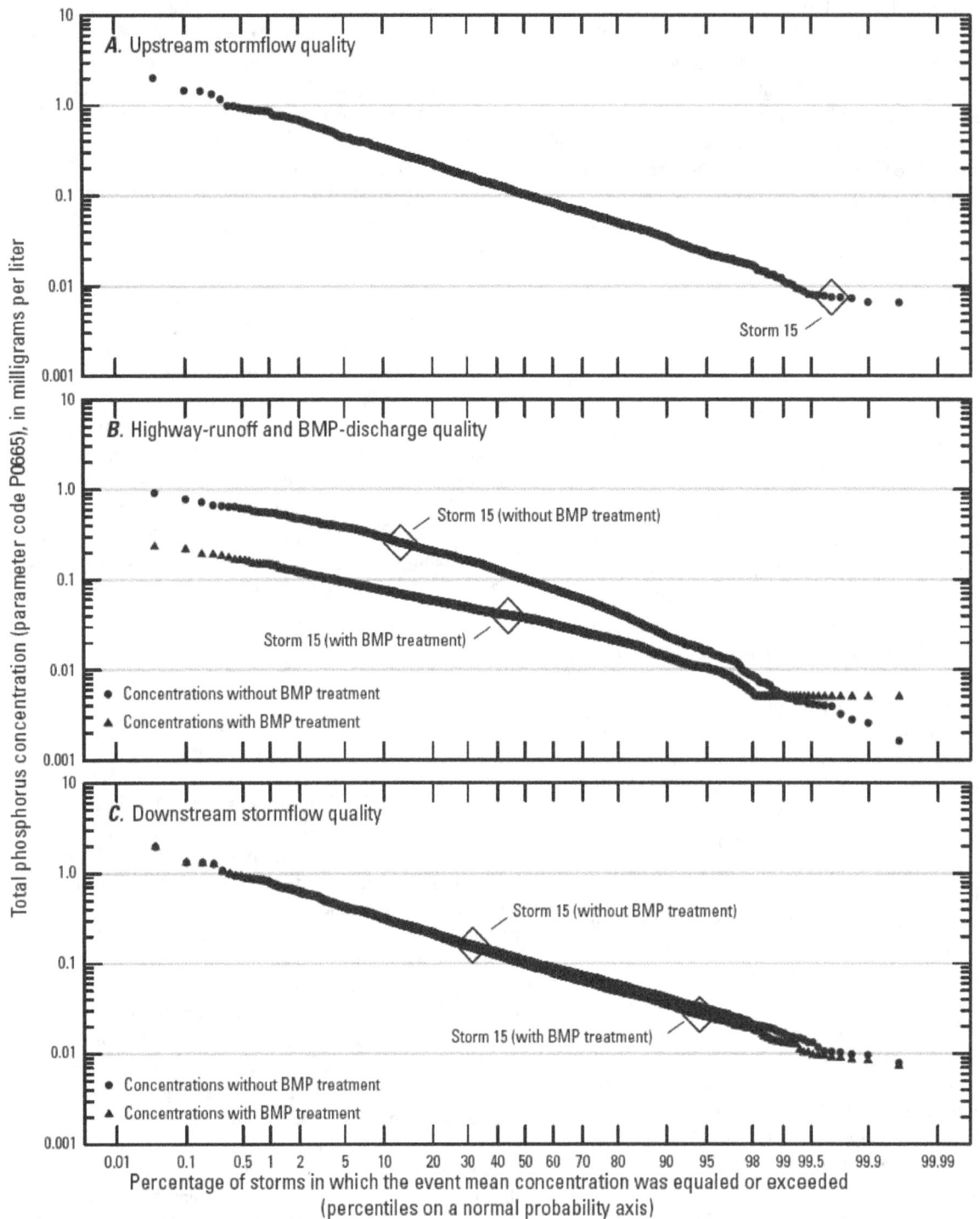

Figure 35.   The stochastic population of total phosphorus concentrations in *A*, upstream stormflow, *B*, highway runoff and discharge from best management practices (BMPs), and *C*, downstream stormflows generated for the I–81 example problem.

the flows and loads concurrent to highway runoff and BMP discharges. In some cases, the duration of highway runoff or BMP discharge may exceed the duration of upstream stormflows. This is more likely if the upstream basin is small, the hydrograph extension by the BMP is large, or both. If the duration of highway runoff or BMP discharge exceeds the duration of upstream stormflows, then SELDM uses the entire upstream stormflow and the prestorm base flow that continues for the duration of highway runoff or BMP discharge after the upstream stormflow has ended. Thus, the sum of annual stormflows from the upstream-basin output may not represent the total annual stormflows and loads for the upstream and downstream sites.

Like the storm-event outputs, the outputs for the highway site and lake basin are associated with plotting-position percentiles that can be used to assess the magnitudes and frequencies of water-quality excursions with or without implementation of the BMPs. The stochastic populations that are generated by this Monte Carlo model can be used to address concerns about the need for probabilistic approaches for defining TMDLs (Smith and others, 2001; Borsuk and

others, 2002; U.S. Environmental Protection Agency, 2002b, 2007a; Bonta and Cleland, 2003; Novotny, 2004; Elshorbagy and others, 2007; Langseth and Brown, 2011). For example, figure 36 shows the plotting positions of 30 annual-load values for highway runoff and BMP discharge. The range of plotting-position percentiles in the annual graphs is smaller than the range in the storm-event graphs (for example, fig. 35) because the number of years and the rank of each year is used to calculate the percentiles rather than the number of storm events. In this example, there are 30 annual values and 1,586 storm-event values.

The load data in figure 36 are plotted on a linear axis because annual loads are the sum of loads for all the independent storm events during each annual-load accounting year. Thus, the annual flows and loads may be better approximated by a normal or Pearson Type III distribution rather than a lognormal or log-Pearson Type III distribution even though the individual storm-event concentrations, flows, and loads may follow a lognormal or log-Pearson Type III distribution. The normal approximation may be sufficient for the loads in figure 36 because the pattern of the data can

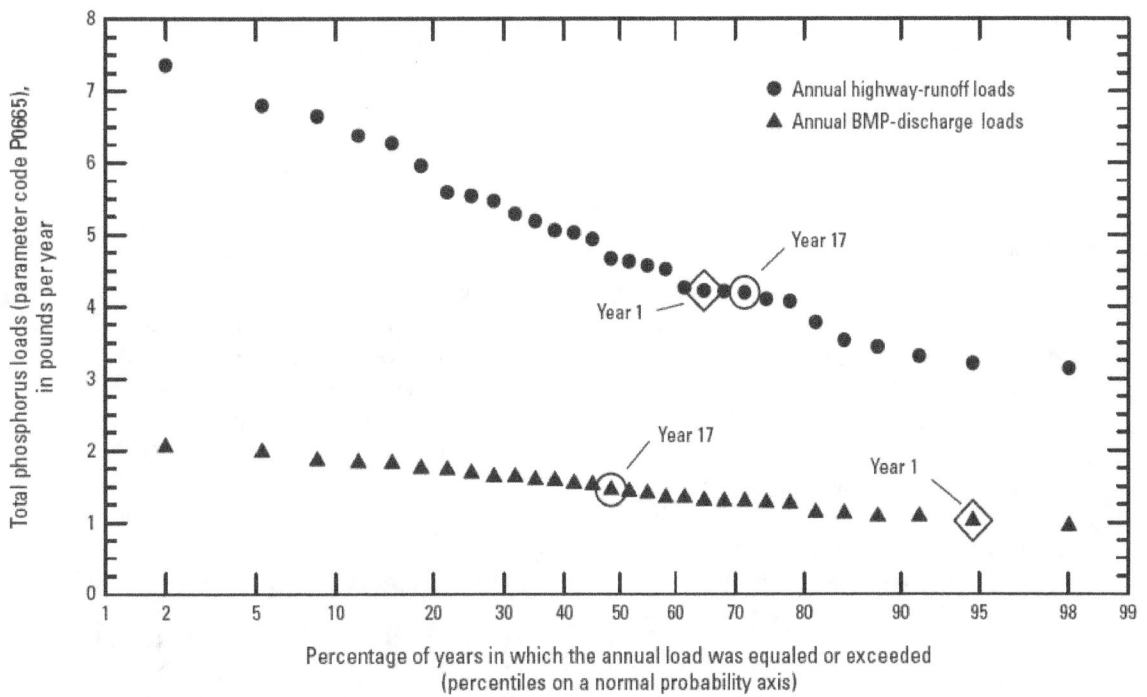

Figure 36. The stochastic population of annual total-phosphorus loads in highway runoff and in discharge from a best management practice (BMP). The highway-runoff values were calculated with statistics from data collected on State Route 2 in Littleton, Massachusetts (Smith and Granato, 2010). The BMP-effluent values were calculated by using the trapezoidal distribution statistics for the Lake Ridge detention pond (fig. 25) and a minimum irreducible concentration of 0.005 milligrams per liter.

be approximated by a straight line with the arithmetic load axis and the probability axis. If the line shows significant curvature (substantial nonzero skew), the Pearson Type III distribution may be a better approximation for annual loads. Although both distributions can produce meaningless negative concentrations, flows, and loads, the COV of annual values commonly is small. To test the applicability of the distribution, multiply -1 by the reciprocal of the COV; the result will be the normal or Pearson Type III variate. The probabilities of loads less than or equal to 0 can be calculated by using commonly available tables of the standard normal or Pearson Type III distributions. Granato (2010) provides electronic tables of the Pearson Type III variates.

SELDM calculates and prints the plotting-position percentiles in the output files, but return periods can be calculated from the outputs if necessary (fig. 37). Unlike the stormflow results, the annual results from SELDM are time based because they account for the total flows and loads from runoff-producing events from the highway site and the total daily loads from the rest of the lake basin. The Cunnane plotting-position formula is used to calculate annual percentiles in the output from SELDM because this formula results in more unbiased values for normal percentiles (Helsel and Hirsch, 2002). If, however, return periods are desired, then it is necessary to convert the plotting positions from Cunnane to Weibull to produce unbiased return-period values (Haan, 1977; Chow and others, 1988; Helsel and Hirsch, 2002). Calculate the return period of annual values from the Cunnane percentiles as

$$Y_R = \frac{1}{\left(\dfrac{P_c(N+0.2)+0.4}{N+1} \right)} \qquad (37)$$

if the data are sorted in descending order and

$$Y_R = \frac{1}{\left(1 - \dfrac{P_c(N+0.2)+0.4}{N+1} \right)} \qquad (38)$$

if the data are sorted in ascending order, where

Y_R is the return period for annual sums, in years;

P_c is the annual Cunnane plotting-position percentile (scaled to the range 0 to 1); and

N is the number of annual-load accounting years generated in a given run of the Monte Carlo model.

Two equations are necessary because recurrence-interval (or return period) plots for annual sums of storm-event concentrations, flows, and loads should be constructed so that larger values have longer return periods. The recurrence intervals are plotted on a logarithmic axis to expand the data over the measurement range. The maximum return period for this example is about 31 years. In this example, on average, the total annual phosphorus loads that are expected to be

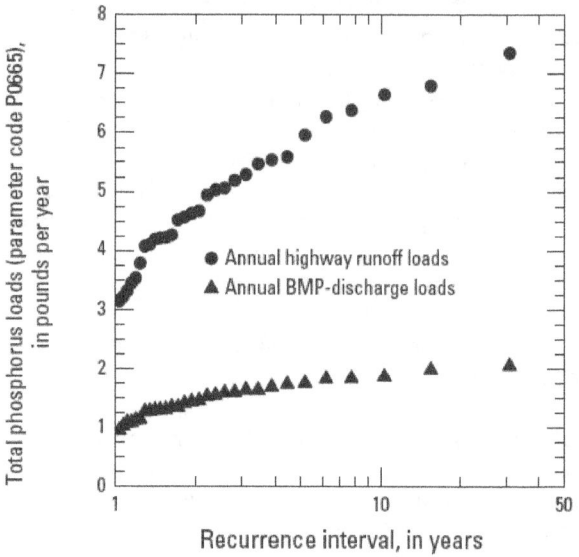

Figure 37. The stochastic population of annual total-phosphorus loads calculated with statistics from highway-runoff data collected on State Route 2 in Littleton, Massachusetts (Smith and Granato, 2010). The values for effluent from best management practices (BMPs) were calculated by using the trapezoidal distribution statistics for the Lake Ridge detention pond (fig. 25) and a minimum irreducible concentration of 0.005 milligrams per liter. The graph shows loads in highway runoff and BMP discharges from a simulated 30-year period plotted by recurrence intervals on a logarithmic axis.

equaled or exceeded once every 3 years (an exceedance percentile of about 31.8 percent) are about 5.29 pounds (lb) for highway runoff and 1.63 lb for BMP effluent; for a recurrence interval of 10 years (an exceedance percentile of about 8.61 percent), the loads are about 6.62 lb for highway runoff and 1.86 lb for BMP effluent. Similar plots and analyses can be made for the other annual highway-output variables and the lake-basin output variables.

Use of the Stochastic Empirical Loading and Dilution Model Interface

SELDM was developed as a Microsoft Access® database software application to facilitate storage, handling, and use of the hydrologic dataset with a simple graphical user interface (GUI). The program's menu-driven GUI uses standard Microsoft Visual Basic for Applications® (VBA) interface controls to facilitate entry, processing, and output of data. Program source code for the analytical techniques is provided within SELDM and also in electronic text files accompanying this report. Program source code that is specific to Microsoft Access® and to the SELDM GUI and that is intended for data handling is provided in the database. This section of the user's manual focuses on installing the model, use of the GUI, and the content and formats of the output files.

Installing the Model

SELDM was designed and implemented using the Microsoft Windows® XP operating system and Microsoft Access® 2003. These versions of Windows® and Access® were used because about 94 percent of the model testers were using the XP operating system, and about 44 percent were using the 2003 version of Microsoft Office® at the time the model was being developed and tested (2010 and 2011). These model testers were from the USGS, FHWA, USEPA, and from state departments of transportation and state regulatory agencies. The model was successfully tested for forward compatibility with the Windows Vista® and Windows 7® operating systems, with Office 2007® and with Office 2010®. A compiled version of the model (SELDMv.01.00a mde) is provided as an independent file and as part of an installation package on the computer media accompanying this report. An uncompiled version of the model (SELDMv.01.00a.mdb) is provided on the computer media accompanying this report to help document the model code and to help facilitate future maintenance of or modifications to the model code. Users may port the uncompiled version to newer versions of Access®, but before doing so, should test the new uncompiled version to ensure that the model-interface code is fully compatible with new Access® reference libraries and data-access models. The compiled (mde) version will run faster than the uncompiled version (mdb) and will be more robust for common use.

The installation package installs the Microsoft Access® runtime program and a compiled version of the model (SELDMv.01.00a mde). Installing the compiled version of the model allows users who do not have a copy of Microsoft Access® to run the model and also ensures that SELDM has all the proper reference libraries to run the model and interface code. Someone having administrative rights on the user's computer must install this software. The folder named "Install" on the computer disk accompanying this report contains a readme.txt file with installation instructions; the installation files are in the same folder. The installation package is a standard Microsoft installation wizard that is likely to be familiar to the user or to the system administrator. Follow the standard choices for software installation to minimize potential problems. The model software and its support files can be uninstalled by using the standard add or remove program wizard found on the control panel for Microsoft Windows®. The installation program creates the directory C:\Program Files\FHWA\SELDM\ and includes the runtime software in the computer's registry. The installation program also creates shortcuts to the compiled version of the model.

Establishing the FHWA-SELDM Output Directory

SELDM is designed to write the model output to a directory named "FHWA-SELDM" on the root drive of the computer. SELDM uses a standard directory to help establish traceability of model results. This output directory must be distinct from the program-files directories used to install SELDM and the accompanying programs that were developed to facilitate analysis of hydrologic and water-quality data (Granato, 2006, 2009, 2010; Granato and Cazenas, 2009; Granato and others, 2009). Running the model without administrator privileges may cause errors if permissions to the root drive are locked. If so, the system administrator must give the user full control (permission to read, write, and modify) of the "FHWA-SELDM" output directory because these permissions are needed to run the model. The system administrator can create the directory manually or can run SELDM to have the model create the directory. If administrator privileges are necessary, the system administrator must set these permissions for the user after SELDM is installed and tested. To test the model, the person installing SELDM can quickly click through the hypothetical example problem that is preloaded in the model.

Using the Source Code or Modifying the Model

The VBA code in SELDM is open source and may be exported and used for other applications designed for private study, scholarship, and research (U.S. Copyright Office, 2000). The VBA subroutines and functions used to implement the Monte Carlo methods, to perform statistical operations, and to manipulate data may also be used in other programs. Some subroutines and functions were developed in VBA from other published works that, if used, must be cited. Some of these sources may require copyright permission for use in commercial software. The rest of the source code for SELDM was developed exclusively by the USGS in cooperation with the FHWA and therefore is in the public domain. If, however, this code is used, the user must cite and annotate it so that any modifications the user has made to the code can be identified.

The SELDM user interface and model code manipulate input and output data by using the database objects. Commonly, data are manipulated by using predefined structured query language (SQL) statements that explicitly manipulate tables and data fields by name according to established database relationships between key fields. The design of the database is documented in appendix 3. Any database-design changes to a compiled version of the model will likely render the model inoperable. Any change in the table names, field names, or database relationships must be accompanied by changes in the source code of an uncompiled version of the model in each place where the VBA code interacts with that portion of the database. These changes must be extensively tested to ensure that the proper working relationships between the database and the code are maintained. Changes to the database design may result in unanticipated consequences in the GUI and in model results. Thus, the database design should not be modified at all unless such modifications are made as part of a comprehensive and systematic redevelopment effort. Any code changes should be thoroughly documented by comment statements in the code. These comments should include information about the author of such changes, the date on which changes were made, and the reason for making the changes.

Navigating the Graphical User Interface

The SELDM user interface has one or more GUI forms that are used to enter four categories of input data, which include documentation, site and region information, hydrologic statistics, and water-quality data (fig. 38). The documentation data include information about the analyst, the project, and the analysis. The site and region data include the highway-site characteristics, the ecoregions, the upstream-basin characteristics, and, if a lake analysis is selected, the lake-basin characteristics. The hydrologic data include precipitation, streamflow, and runoff-coefficient statistics. The water-quality data include highway-runoff-quality statistics, upstream-water-quality statistics, downstream-water-quality definitions, and BMP-performance statistics. There also is a GUI form for running the model and accessing the distinct set of output files. The SELDM interface is designed to populate the database with data and statistics for the analysis and to specify index variables that are used by the program to query the database when SELDM is run. It is necessary to step through the input forms (appendix 4) each time an analysis is run.

To facilitate use of the GUI, most of the forms display a similar layout and use the same controls for entering data and the same options for processing them. Individual forms that require exceptions to these standards are noted where applicable. An example of the standard components and appearance of most forms is provided by the Analysis Identification Form (fig. 39). The first line of the form, at the upper left, is the title that describes the form's contents. At the upper right, an Information button opens a form that provides text explaining the purpose and scope of the current form. The selection combobox near the top of the form lets users choose from previously entered data. Figure 39A displays the title of the selection combobox as Select Analysis, and it displays Example Analysis (I–81) as the current selection. To select the desired data, left-click on the down arrow in the combobox, and left-click again on the desired choice. On most forms, the default selection for the previously entered data is the last data entered.

The next component on most forms is the option-selection frame (fig. 39), which controls the appearance and operation of the form. Many of the forms have the option to Select Current, Edit Current, and Enter New data. Some forms also have the Copy Current option. The Select Current option

is the default choice when each form opens. The data-entry fields (beneath these options on the form) are disabled and locked (as their grayed-out fields indicate): the data are visible, but cannot be changed when the form is in the Select Current mode (fig. 39A); this prevents the inadvertent modification of data. Changing the selection in the combobox will change the data in the fields, but the data-entry fields will remain locked and disabled. Clicking the Edit Current option unlocks and enables the fields (fig. 39B); the color of the field will change from gray to white, and you will be able to click in the field and modify or enter data. Although the Select Current option is the default option in SELDM, the figures showing the GUI were made using the Edit Current option selected in figure 39B to better show the contents of the available fields. If you click the Enter New option, this selection unlocks, enables, and clears the data-entry fields, and it also disables the selection combobox. Some forms have a Copy Current option, which unlocks and enables the data-entry fields and the Copy button. The Copy Current option can be used create a new analysis with the same input values as the original. Copying an analysis and modifying selected variables can be used for doing a sensitivity analyses or for scenario testing.

The command buttons on the bottom of the form (fig. 39) let you finalize data entry and navigate between forms. Most forms have an Exit SELDM button to close the form and exit the database application. Most forms also have an Accept Updates button to check the inputs and save the data to the SELDM database. Most forms have a Proceed button to go to the next form, and a Go Back button to return to the previous form.

Choosing different options in the option-selection frame at the top of the form affects the status of the command buttons on the bottom of the form. Clicking the Edit Current or Enter New selections in the option-selection frame enables the Accept Updates button and disables the Proceed button so that you can save changes to data. If Copy is an available option clicking on this option will disable the Proceed button and enable the Copy button on the bottom of the form. You can discard any changes by clicking the Select Current option before clicking the Accept Updates or Copy buttons. Clicking the Select Current option will enable the Proceed button and disable the Accept Updates and the Copy buttons.

On most forms, the data-entry fields are arranged on a group of tabbed pages (fig. 39). The top tabbed page commonly is for entering introductory information. Underlying tabbed pages are for associated data. On the Analysis Identification Form there is one tabbed page for associated data labeled Analysis Options. The last tabbed page on seven of the data-entry forms has a large text field designed to let you record explanatory information (for example, citing a report as the source of your input data). On the Analysis Identification Form, this tab is labeled Analysis Description.

Mandatory fields on all the forms are marked by an asterisk. Mandatory data in these fields are necessary for populating the interface and running the model. The titles of

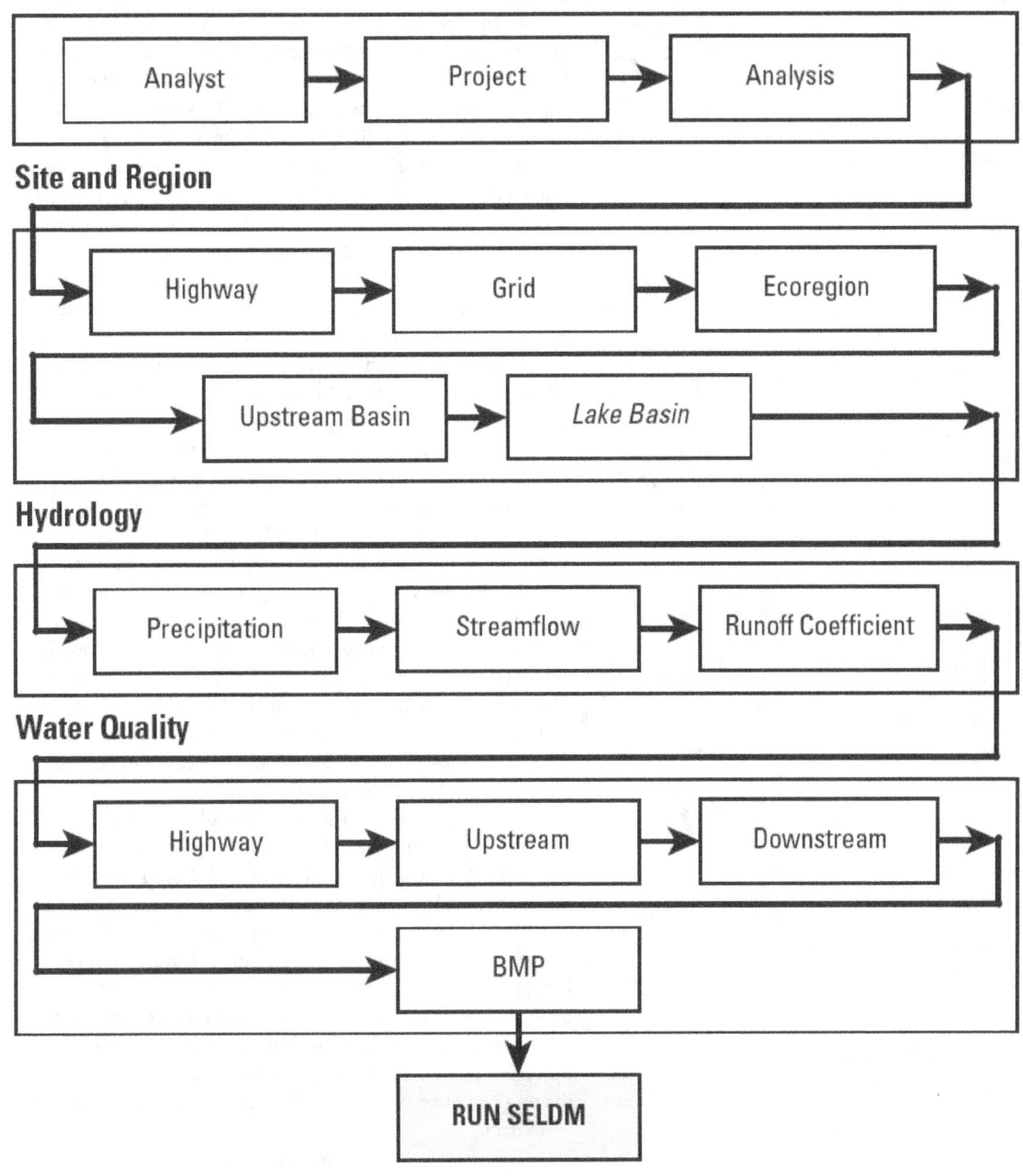

Figure 38. The flowchart for entering input data and for running an analysis. The component "Lake Basin" is italicized because this form is not used in all analyses. BMP, best management practice.

A

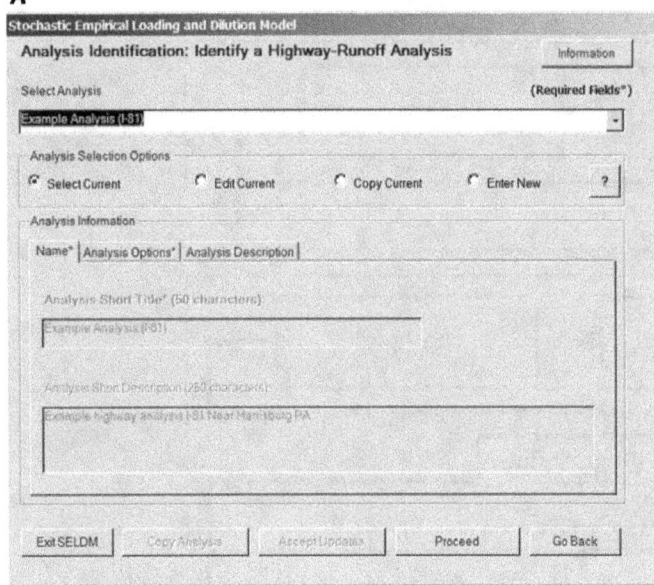

B

Figure 39.   The Analysis Identification Form in the *A,* select current and *B,* edit current modes showing the basic components of many SELDM data input forms.

data-input tabs that contain mandatory input fields also are marked by an asterisk.

Each type of data is assigned a short, meaningful, and distinct name that is used in the selection combobox on each form. The short name must be of limited length so it can be displayed properly in the combobox. The short name should be meaningful and distinctive to facilitate selection using the combobox. Only primary ASCII characters, including letters (uppercase and lowercase), numbers, dashes, underscores, and parentheses, are permitted for short names. The short names are printed to the output files with the other data.

Most forms have one or more context-sensitive help buttons denoted with a question mark "?" (fig. 39). These buttons are placed adjacent to related form controls and fields. Left-clicking these context-sensitive help buttons with the mouse cursor launches an Information Form or a message box with relevant information.

All the forms use two Microsoft® form-control help features—the tool-tip text and status-bar text—which provide information about each control on the form. These help features summarize the content of each data-entry feature. Tool-tip text messages appear as popup messages when you position the mouse cursor over any control on the form for a few seconds. Status-bar text messages appear in the lower-left corner of the Microsoft Access® application window. The content of each of these features is limited to 255 characters. The amount of status-bar text that can be displayed, however, may be limited by your choice of a Windows font size and the way you resize the Access window, because Access can display only as much text as will fit in the status bar.

The two methods for navigating within a form are using the mouse and using the Tab key. These methods can be used in combination. Using the mouse to click from feature to feature is the most commonly used method because you can click on any control in any order with this method. Using the Tab key to step through a form's controls is useful but less flexible because you must follow the predetermined tab order, and you are limited to fields within the active tabbed page. You must click on the page with the mouse before tabbing through the fields on that page. However, the tab key can be the most efficient way to step through a series of text fields to enter data.

This user's manual is focused on modeling runoff-quality rather than using the SELDM GUI. The SELDM GUI is designed to lead you through the flow chart shown in figure 38 so that you may enter or edit the data and statistics that are described in the theory and implementation section of this report. Detailed information about each form is described in detail in appendix 4.

Model-Output Files

The results of each SELDM analysis are written to 5 to 10 output files, depending on the options you select during the analysis-specification process (table 7). The five output files that are created for every model run are the output documentation, highway-runoff quality, annual highway runoff, precipitation events, and stormflow file. The output options you chose on the Analysis Identification Form (fig. 39) determine which of the remaining five output files are also created. If the Stream Basin or Stream and Lake Basin output options are selected, then the prestorm streamflow and dilution factor files also are created. If these same two output options are selected and, in addition, one or more downstream water-quality pairs are defined by using the water-quality menu (appendix 4, fig. 4–20), then the upstream water-quality and downstream water-quality output files also are created by SELDM. If the Stream and Lake Basin Output or Lake Basin Output option is selected (appendix 4, fig. 4–6), and one or more downstream water-quality pairs are defined by using the water-quality menu (appendix 4, fig. 4–20), then the Lake Analysis output file is created when the Lake Basin Analysis is run by clicking on the Run The Lake Package button on the Run SELDM form (appendix 4, fig. 4–29). The name of each output file is the Analysis short name with the associated output-file suffix (table 7).

The output files are written as tab-delimited ASCII text files in a relational-database (RDB) format that can be imported into many software packages (Manis and others, 1988). The SELDM RDB file format begins with comment lines that describe the content of the file, the analysis-specification options, and the output-value columns. A pound symbol (#) in the first character location of each line denotes these comment lines. The comments are followed by a line of variable names for each type of output value in the file. These documentation and format lines in these files are essential for documenting model outputs in a meaningful and efficient format. The remaining output lines in the RDB files contain the numerical results of the analysis. The tab-delimited format of the numerical-result output tables is designed to be copied and pasted into a spreadsheet or graphing package for presentation of results or for further analysis. If an output file includes more than one block of numerical outputs, then each block of numerical outputs will be preceded by a block of comment lines that explain the specific outputs. Example output files are provided in the directory named "ExampleOutput" on the CD–ROM containing this manual.

All the output files that contain numerical simulation results have similar heading information. The first line of text in all the output files identifies the contents of the file as SELDM output and lists the date and time (in hours, minutes, and seconds) that the model was run. This date stamp can be used with the file-system date stamp to verify that the contents of the file are unaltered SELDM outputs. The date stamp is the same for all output files except the lake-analysis output file, on which the stamp indicates the time that the lake analysis was

Table 7. Output files from the Stochastic Empirical Loading and Dilution Model (SELDM) associated with the stream-basin, stream- and lake-basin, and lake-basin analyses.

[Type 1, analysis with downstream-pair definitions; Type 2, analysis without downstream-pair definitions; X, has the specified type of output; Y, has the highway-stormflow output, but not the upstream-stormflow output; --, does not have the specified type of output]

File		Stream-basin analysis		Stream- and lake-basin analysis		Lake-basin analysis	
Type	Suffix	Type 1	Type 2	Type 1	Type 2	Type 1	Type 2
Output documentation file	Out.txt	X	X	X	X	X	X
Precipitation events	-PE.txt	X	X	X	X	X	X
Prestorm streamflow	-PS.txt	X	X	X	X	--	--
Stormflows	-SF.txt	X	X	X	X	Y	Y
Dilution factor	-DF.txt	X	X	X	X	--	--
Highway runoff quality	-HQ.txt	X	X	X	X	X	X
Annual highway runoff	-Annual.txt	X	X	X	X	X	X
Upstream water quality	-UQ.txt	X	--	X	--	--	--
Downstream water quality	-DQ.txt	X	--	X	--	--	--
Lake analysis	-Lake.txt	--	--	X	--	X	--

run. The second line of text is a title that indicates the content of each output file. The next block of text lists the version number, the version date, and the citation for the version of SELDM being used. The fourth block of text briefly describes the file contents and identifies the output-documentation file (Out.txt) as the source for further information and disclaimers. The fifth block of text briefly describes the annual and storm sequences as a random series of values that are generated regardless of sequence or seasonality. The sixth block of text explains the fact that the plotting position for each variable reflects the results of sorting the values of the individual variables rather than assigning a plotting position to all the values generated for a specific storm. The seventh block of text is the analysis information, which includes the short title, short description (if it is specified), type of analysis, status of analysis, and full description (if it is specified). The remainder of the information in these output files is context sensitive, because the output depends on the type of information in the output file.

The first numerical column of each storm-event output file contains the storm-event sequence number. Individual storm events are listed in the order in which they were generated by the model and are identified by sequence number. SELDM generates each storm randomly; there is no serial correlation. The order of storms does not reflect seasonal patterns. Flows and loads are not propagated from storm to storm. Different highway-runoff constituents generated for each storm are not correlated unless a dependent water-quality relationship is specified. Different upstream constituents generated for each storm are not correlated unless a dependent water-quality relationship is specified or the constituents are generated from stormflow data by using a water-quality transport curve. The storm-event sequence numbers are provided to facilitate generation of paired-value scatterplots and examination and analysis of the output to check and verify the results of model calculations.

The second numerical column of each storm-event output file and the first numerical column of the annual highway-runoff and lake-analysis output files contain the year number. The year number denotes an annual-load accounting year rather than a historical time series. Annual-load accounting years are just random collections of events generated with storm durations and inter-event times that sum to a value that is less than or equal to one year (8,760 to 8,784 hours).

The remaining output columns consist of paired plotting-position and hydrologic values. The plotting-position values are calculated by sorting the associated hydrologic variable and calculating the plotting position as a fraction or a percent. Options for sorting and calculating plotting positions are selected by using the Run SELDM form (appendix 4, fig. 4–29). Plotting positions for storm-event values are calculated by using the total number of events generated. Plotting positions for the annual values are calculated by using the total number of annual-load accounting years generated.

Output-Documentation File

The SELDM output-documentation file (Out.txt) provides basic information about the analysis and is, in effect, an annotated outline for the selected output files created for a given analysis (table 7). This Out.txt file lists the name of each output file with a description of the type of information in each output file. This file has 15 blocks of text documenting inputs to the model, standard model disclaimers, and basic information about the results of the analysis. This file also documents the properties of the BMP selected for highway-runoff control. If a BMP is selected, the short name for the selected BMP is listed with other information; otherwise, the text "The BMP information is not defined" will appear in its place. The full name and description for the selected BMP will be included if they are specified. Depending on user inputs, the BMP flow-reduction specifications, hydrograph-extension specifications, and water-quality-modification definitions will be listed in this output file. Detailed information about the BMP performance options is described in the section of this report titled "Runoff Modification by Best Management Practices (BMPs)."

Precipitation-Event Output File

The Precipitation-Event output file (-PE.txt) includes information about precipitation selections that were input to the model and numerical outputs for each storm event in the random sample that SELDM generates. This file documents the precipitation dataset that was used in the analysis, the selection used to calculate statistics, and the calculation method. Unless the user-defined option was selected on the Synoptic Storm-Event Precipitation Statistics form (appendix 4, fig. 4–13), this file also contains the list of hourly-precipitation data stations that were used to calculate the event statistics. The numerical outputs include a table listing the interval between storm-event midpoints, the precipitation volume of the each generated storm event, the duration of each generated storm event, and the paired plotting-position values for each variable. The approach for defining and generating precipitation events is briefly described in the section of this report titled "Storm-Event Characteristics." Granato (2010) provides a detailed discussion of methods for calculating precipitation statistics for use with SELDM.

Prestorm-Streamflow Output File

The prestorm-streamflow output file (-PS.txt) includes information about the upstream basin and streamflow statistics that were input to the model and numerical outputs for the upstream prestorm streamflows for each storm event in the random sample that SELDM generates. This file documents the streamflow dataset that was used in the analysis, the selection used to calculate statistics, and the calculation method. Unless the user-defined option was selected on the Streamflow Statistics form (appendix 4, fig. 4–16), this file also contains the list of streamgages that were used to calculate the flow statistics. The numerical outputs include a table of upstream-flow rates for each storm and the paired plotting-position values. The prestorm streamflow is used as an approximation for base flows during each modeled runoff event. The approach for defining and generating prestorm streamflow is briefly described in the section of this report titled "Prestorm Streamflow Volumes." Granato (2010) provides a detailed discussion of methods for calculating streamflow statistics to estimate prestorm streamflow for SELDM.

Stormflow Output File

The stormflow output file (-SF.txt) includes information about the highway site, the upstream basin, stormflow selections, and numerical outputs for the components of the stormflow analysis for each storm event in the random sample that SELDM generates. This file documents the input runoff-coefficient statistics, BMP hydrograph-extension statistics, and BMP flow-reduction statistics that were used in the analysis. The components of the stormflow analysis for each modeled runoff event include runoff coefficients, storm-event durations, and discharges from the highway site and the upstream basin for each storm event in the random sample generated by SELDM. The stormflow output file contains four tables of numerical data that include runoff coefficients, storm-event durations, highway-runoff and BMP discharges, upstream stormflow, and the paired plotting-position values for each variable. The runoff-coefficient table has runoff coefficients for the highway and the upstream basin. The storm-event-duration table has durations for highway runoff, BMP discharge, and the total upstream streamflow for a storm event. The highway-runoff and BMP-discharge table has the volume of runoff from the highway and the volume discharged from the BMP; these values may be equal if there is no BMP, or if the selected BMP does not produce discernible reductions in flow. The upstream-stormflow table has the upstream runoff, the total upstream flow, the total upstream flow concurrent to the highway-runoff duration, and the total upstream flow concurrent to the BMP-discharge duration. The total upstream flow concurrent to the highway-runoff duration and the total upstream flow concurrent to the BMP-discharge duration may be equal to the highway-runoff duration if there is no BMP, or if the selected BMP does not produce discernible extensions to the highway-runoff durations. The approach for defining and generating stormflows from the highway site and upstream basin are briefly described in the section of this report titled "Stormflow." Granato (2010) provides a detailed discussion of methods for calculating stormflow statistics for SELDM. The approach for defining and generating flow-modification statistics is briefly described in the section of this report titled "Runoff Modification by Best Management Practices (BMPs)."

Dilution-Factor Output File

The dilution-factor output file (-DF.txt) includes information about the highway site, the upstream basin, and numerical outputs for the dilution factor for each storm event in the random sample that SELDM generates. The dilution factor is the total highway discharge divided by the concurrent downstream stormflow for each storm event in the random sample generated by SELDM. The dilution-factor output provides a simple assessment of the potential for adverse effects from highway runoff. A dilution factor of 1 indicates that all of the downstream flow is highway runoff, and a dilution factor near 0 indicates that highway runoff is a negligible portion of the downstream flow. The numerical outputs include the highway-runoff and BMP-discharge dilution factors for each storm event and the paired plotting-position values for each variable. The highway runoff and BMP discharge dilution factors will be equal if there is no BMP, or if the selected BMP does not produce discernible flow reductions. More detailed information about the calculation and interpretation of dilution factors is described in the section of this report titled "Dilution Factors." SELDM does not calculate the dilution factors at the end of a mixing zone downstream of the highway outfall, but it can be used to estimate these factors if the upstream-basin area is adjusted. Detailed information about this downstream flow-accretion analysis method is described in the section of this report titled "Mixing."

Highway-Runoff-Quality Output File

The highway-runoff-quality output file (-HQ.txt) includes information about the highway site, input runoff-quality specifications, and numerical outputs for each selected highway-runoff constituent for each storm event in the random sample generated by SELDM. Each highway-runoff-quality output table is preceded by constituent-definition information including the short name of the constituent, the USEPA parameter code (PCode), the parameter name (which includes the water-quality units), the method for generating the output, the random or dependent water-quality statistics, and the water-quality abstract (if this information is entered). The runoff-quality output table for each selected constituent includes the event mean concentration of highway runoff, the total highway-runoff load, the event mean concentration of discharge to the stream, the total runoff load discharged to the stream, and the paired plotting-position values for each variable. The highway-runoff and discharge concentrations and loads will be equal if there is no BMP, or if the selected BMP does not produce discernible reductions in constituent concentrations. The approach for stochastic generation of highway-runoff-quality constituents is described in the section of this report titled "Highway and Upstream Stormwater Concentrations and Loads." Granato and Cazenas (2009) provide a detailed discussion of methods for calculating the input highway-runoff-quality statistics.

Annual Highway-Runoff Output File

The annual highway-runoff quality output file (-Annual.txt) includes input information about the highway site, input runoff-quality specifications, and annual numerical outputs generated by SELDM for each annual-load accounting year. These annual numerical outputs include total annual precipitation volumes, flows and loads of highway runoff, and flows and loads of BMP discharge for each selected constituent. Output values and the associated plotting-position values are produced for each annual-load accounting year in the random sample generated by SELDM. Each highway-runoff-quality output table is preceded by constituent-definition information including the short name of the constituent, the PCode, the parameter name (which includes the water-quality units), the method for generating the output, the random or dependent water-quality statistics, and the water-quality abstract (if this information is entered). This file also contains the total annual highway-runoff flow and BMP discharge in volumetric units (cubic feet or cubic meters) and in normalized units (in watershed inches or watershed centimeters for the highway drainage area) for each selected runoff-quality constituent. The annual highway-runoff-quality output file also contains the total annual highway-runoff and BMP-discharge loads for each selected constituent. These flows and loads will be equal if there is no BMP, or if the selected BMP does not produce discernible reductions in runoff flows or constituent concentrations. The annual flows and loads for each annual-load accounting year are the sums of each storm-event value assigned to that year, but the annual flows and loads may not equal the sum of individual values in the highway-runoff-quality output file because of rounding to the specified number of significant digits. More detailed information about the calculation and interpretation of annual flows and loads is described in the section of this report titled "Interpreting the Results of an Analysis."

Upstream Runoff-Quality Output File

The upstream runoff-quality output file (-UQ.txt) includes information about the upstream basin, input stormflow-quality specifications, and numerical outputs for each selected upstream water-quality constituent for each storm event in the random sample generated by SELDM. Each upstream runoff-quality output table is preceded by constituent-definition information including the short name of the constituent; the PCode; the parameter name (which includes the water-quality units); the method for generating the output; the random, dependent, or transport-curve statistics; and the water-quality abstract (if this information is entered). The upstream runoff-quality output table for each selected constituent includes the event mean concentrations of upstream runoff and loads concurrent to the highway runoff, the event mean concentrations of upstream runoff and loads concurrent to the highway BMP discharge, and the paired plotting-position

values for each variable. These upstream-runoff concentrations and loads will be equal if there is no BMP, or if the selected BMP does not produce discernible reductions in flows and in constituent concentrations. The approach for stochastic generation of upstream runoff-quality constituents is described in the section of this report titled "Highway and Upstream Stormwater Concentrations and Loads." Granato and others (2009) provide a detailed discussion of methods for calculating the input upstream stormflow-quality statistics.

Downstream Runoff-Quality Output File

The downstream runoff-quality output file (-DQ.txt) includes information about the highway site, the upstream basin, input stormflow-quality specifications, and numerical outputs for each selected downstream water-quality pair for each storm event in the random sample generated by SELDM. Each downstream-runoff-quality output table is preceded by constituent-definition information including the short name of the constituent; the PCode; the parameter name (which includes the water-quality units); the method for generating the output; the random, dependent, or transport-curve statistics; and the water-quality abstract (if this information is entered). This information is provided for both the highway-runoff and the upstream-basin constituents that are used to define a pair. The downstream runoff-quality output table includes the fully mixed downstream event mean concentration, the associated adverse-effect concentration, the downstream stormflow, the downstream load for each selected constituent pair, and the plotting-position values for each variable. These concentrations and loads are calculated for the storm volumes that are concurrent to the highway runoff or to the BMP discharge as defined for the selected water-quality pair. If the option for using BMP flow and water-quality treatment is selected when a downstream pair is defined (appendix 4, fig. 4–24), then the downstream water-quality variables will be labeled "concurrent to the highway BMP discharge." If the BMP option is not selected, the downstream water-quality variables will be labeled "concurrent to the highway runoff." The adverse-effect concentrations will equal the downstream concentration if the Adverse-Effect Concentration Statistics option is not selected (appendix 4, fig. 4–24). The approach for stochastic generation of downstream water-quality pairs, including the adverse-effect concentration, is described in the section of this report titled "Downstream Stormwater Concentrations and Loads."

Lake-Analysis Output File

The lake-analysis output file (-Lake.txt) includes information about the highway site, the lake basin, input lake-analysis specifications, and annual numerical outputs generated by SELDM for each annual-load accounting year. These annual numerical outputs include total annual flows, total annual loads, the annual detention time, the average annual lake concentration and the associated plotting-position values for each output value. The annual flows and loads are calculated for the highway site and the whole lake basin with and without the highway-site contributions. Each lake-analysis output table is preceded by constituent-definition information, including the short name of the constituent, the PCode, the parameter name (which includes the water-quality units), the method for generating the output, the random or dependent water-quality statistics, and the water-quality abstract (if this information is entered). The lake-analysis output table for each selected constituent includes the flows, loads, concentrations, and lake detention times associated with that constituent definition. The annual flows and loads from the highway for each annual-load accounting year are the sums of the storm-event values assigned to that year, but the annual flows and loads may not equal the sums of individual values in the highway-runoff quality output files because of rounding to the specified number of significant digits. The annual flows and loads from the lake basin for each annual-load accounting year are the sums of the 365 or 366 daily flows and loads generated for the lake-basin analysis. Both dry-weather and storm-event flows and loads from the lake basin are included because dry-weather flows and loads may represent a substantial proportion of the annual totals. More detailed information about the calculation and interpretation of annual lake-basin flows and loads is described in the section of this report titled "Lake-Basin Analysis."

Each variable in the output file is associated with a plotting-position variable from the stochastic analysis that indicates the probability of exceedance of that variable. As with the main SELDM module, the user can specify whether the plotting positions are expressed as fractions or as percentiles and are assigned to output values in ascending or descending order. The sorting order affects the interpretation of the plotting positions. If an output variable is sorted in ascending order, which is the default, then the plotting position indicates the magnitude of values less than or equal to the value associated with the selected plotting position. For example, if the 20th-percentile value (or a plotting position of 0.2) is equal to 10, then there is a 20-percent chance that the average annual value will be less than or equal to 10 and an 80-percent chance that it will be greater than 10 for any given year. If the output variable is sorted in ascending order, a value of 10 would be assigned to the 80th percentile (or a plotting position of 0.8). The SELDM lake module does not output target concentrations or loads because such values commonly are context sensitive, or they are established for a political jurisdiction. The plotting-position output can be used to assess the risk for exceeding any target value. The SELDM lake-basin module can be run to simulate the effects of different mitigation measures, and the plotting positions can be used to estimate the potential effects of such measures on the probability of exceeding target values.

Summary

The Stochastic Empirical Loading and Dilution Model (SELDM) is a tool that can be used to transform disparate and complex scientific data into meaningful information about the risk for adverse effects of runoff on receiving waters, the potential need for mitigation measures, and the potential effectiveness of such management measures for reducing these risks. The U.S. Geological Survey developed SELDM in cooperation with the Federal Highway Administration to generate planning-level estimates of event mean concentrations, flows, and loads. This information is needed for highway engineers, regulators, and decisionmakers to evaluate the potential effects of highway runoff on receiving waters and, if necessary, to help guide the choice of potential mitigation strategies. SELDM uses information and data about a highway site, a receiving-water basin, precipitation events, stormflow, water quality, and the performance of mitigation measures to produce a stochastic population of runoff-quality variables. SELDM provides input statistics for precipitation, prestorm flow, runoff coefficients, and selected water-quality constituents derived from National datasets. These input statistics may be selected on the basis of the latitude and longitude of the site of interest and the characteristics of the highway site and the upstream basin. The user also may derive and input statistics that are specific to a given site or a given area for each variable.

SELDM is a stochastic model because it uses Monte Carlo methods to produce the random combinations of input-variable values needed to generate the stochastic population of values for each component variable. SELDM uses concentrations, flows, and loads from a highway site and a receiving-water basin to calculate the dilution of runoff in the receiving waters and the resulting downstream event mean concentrations and annual lake concentrations. Results are ranked, and plotting positions are calculated, to indicate the risk that runoff concentrations, flows, and loads may cause adverse effects on receiving waters by storm and by year. Unlike deterministic hydrologic models, SELDM is not calibrated by changing values of input variables to match a historical record of values. Instead, input values for SELDM are based on site characteristics and representative statistics for each hydrologic variable. Thus, SELDM is an empirical model based on data and statistics rather than theoretical physiochemical equations.

SELDM is a lumped parameter model because the highway site, the upstream basin, and the lake basin each are represented as single homogenous units. Each of these source areas is represented by average basin properties, and results from SELDM are calculated as point estimates at the site of interest. For example, highway-runoff results are produced for the highway site and the outlet of a BMP, if a BMP is specified. Upstream-basin results are produced for the point at which the highway runoff enters the stream. Average annual lake-basin results are produced for the entire (well mixed) lake. Use of the lumped parameter approach provides a number of advantages over distributed modeling approaches. First, use of the lumped parameter approach facilitates rapid specification of model parameters to develop planning-level models. Second, this approach also is representative of the detail available in most available datasets. For example, available datasets for highway runoff and BMP performance commonly are limited to a small number of representative sites. Similarly, watershed studies for stream- or lake-quality monitoring commonly are based on data collected at sites that represent multiple land covers and different tributaries in a single drainage basin. Third, this approach allows for parsimony in the required inputs to and outputs from the model. Fourth, this approach allows flexibility in the use of SELDM. For example, because the highway site is defined by a few hydraulic parameters, the site definition also can be used to model runoff from various land covers by using the appropriate impervious fraction and representative runoff concentrations. In addition, the lumped parameter highway-site definition can be used to form conceptual runoff models with a unit drainage area and an impervious fraction to calculate annual loads for different highway sites or land covers. For example, if data for constituent concentrations representing runoff from different road classes or land covers are generated for a generic one-acre site, then the estimated annual loads per unit area can be used with a geographic information system (GIS) program to estimate total loads from the sum of different contributing areas in a watershed.

This report is a user's manual for the model. It provides information about the theory and implementation of the model (including an appendix describing Monte Carlo methods), deriving model inputs (including an appendix describing the basin properties needed to characterize the highway site and the upstream basin), and interpreting model outputs. It also provides a detailed discussion of the graphical user interface and the format of output files.

References Cited

Abramowitz, Milton, and Stegun, I.A., eds., 1964, Handbook of mathematical functions with formulas, graphs, and mathematical tables: U.S. Department of Commerce, National Bureau of Standards, Applied Mathematics Series, v. 55, 1,046 p.

Adams, B.J., Fraser, H.G., Howard, C.D.D., and Hanafy, M.S., 1986, Meteorological data analysis for drainage system design: Journal of Environmental Engineering, v. 112, no. 5, p. 827–848.

Adams, B.J., and Papa, Fabian, 2000, Urban stormwater management planning with analytical probabilistic models: New York, John Wiley & Sons, Inc., 358 p.

Allen, H.E., and Hansen, D.J., 1996, The importance of trace metal speciation to water quality criteria: Water Environment Research, v. 68, no. 1, p. 42–54.

Allison, J.D., and Allison, T.L., 2005, Partition coefficients for metals in surface water, soil, and waste: U.S. Environmental Protection Agency Report EPA/600/R–05/074, 93 p.

Allison, J.D., Brown, D.S., and Novo-Gradac, K.J., 1990, MINTEQA2/PRODEFA2, A geochemical assessment model for environmental systems, v. 3.0—User's Manual: U.S. Environmental Protection Agency Report EPA/600/3–91/021, variously paged.

American Society for Testing and Materials, 2009, D3977–97R07, Standard test method for determining sediment concentration in water samples—Annual book of standards: Water and Environmental Technology 2009, v. 11.02.

Athayde, D.N., Shelley, P.E., Driscoll, E.D., and Boyd, G.B., 1983, Results of the nationwide urban runoff program, v. I—Final report: U.S. Environmental Protection Agency Water Planning Division Report, WH–554, 186 p.

Bachmann, R.W., 1981, Prediction of total nitrogen in lakes and reservoirs: U.S. Environmental Protection Agency Report EPA 440/5–81–010, p. 320–324.

Back, W.E., Boles, W.W., and Fry, G.T., 2000, Defining triangular probability distributions from historical cost data: Journal of Construction Engineering, v. 126, no. 1, p. 29–37.

Bank, F.G., Kerri, K.D., Young, G.K., and Stein, Stuart, 1996, National evaluation of water quality issues for highway planning: Federal Highway Administration, accessed May 26, 2012, at http://environment.fhwa.dot.gov/ecosystems/wet_wqnatevl.asp.

Barnwell, T.O., Jr., and Krenkel, P.A., 1982, Use of water quality models in management decision making: Water Science and Technology, v. 14, no. 9–11, p. 1095–1107.

Barrett, M.E., 2005, Performance comparison of structural stormwater best management practices: Water Environment Research, v. 77, no. 6, p. 78–86.

Barrett, M.E., 2008, Comparison of BMP performance using the International BMP Database: Journal of Irrigation and Drainage Engineering, v. 134, no. 5, p. 556–561.

Bartsch, A.F., and Gakstatter, J.H., 1978, Management decisions for lake systems on a survey of trophic status, limiting nutrients, and nutrient loadings, in American-Soviet Symposium on Use of Mathematical Models to Optimize Water Quality Management: Gulf Breeze, Fla., U.S. Environmental Protection Agency Report EPA 600/9–78–024, p. 372–394.

Benoit, G., 1994, Research communications—Clean technique measurement of Pb, Ag, and Cd in freshwater—A redefinition of metal pollution: Environmental Science and Technology, v. 28, no. 11, p. 1987–1991.

Benoit, G., Hunter, K.S., and Rozan, T.F., 1997, Sources of trace metal contamination artifacts during collection, handling, and analysis of freshwaters: Analytical Chemistry, v. 69, no. 6, p. 1006–1011.

Bent, G.C., Gray, J.R., Smith, K.P., and Glysson, G.D., 2003, A synopsis of technical issues for monitoring sediment in highway and urban runoff, in Granato, G.E., Zenone, Chester, and Cazenas, P.A., eds., National highway runoff water-quality data and methodology synthesis, v. I —Technical issues for monitoring highway runoff and urban stormwater: Federal Highway Administration Report FHWA–EP–03–054, p. 111–163.

Bhavsar, S.P., Gandhi, Nilima, Diamond, M.L., Lock, A.S., Spiers, Graeme, and de la Torre, M.C.A., 2008, Effects of estimates from different geochemical models on metal fate predicted by coupled speciation-fate models: Environmental Toxicology and Chemistry, v. 27, no. 5, p. 1020–1030.

Biesecker, J.E., and Leifeste, D.K., 1975, Water quality of hydrologic bench marks—An indicator of water quality in the natural environment: U.S. Geological Survey Circular 460–E, 21 p.

Bishop, G.D., and Church, M.R., 1995, Mapping long-term regional runoff in the eastern United States using automated approaches: Journal of Hydrology, v. 169, p. 189–207.

Bobee, Bernard, and Ashkar, Fahim, 1991, The gamma family and derived distributions applied in hydrology: Littleton, Colo., Water Resources Publications, 203 p.

Bonta, J.V., and Cleland, Bruce, 2003, Incorporating natural variability, uncertainty, and risk into water-quality evaluations using duration curves: Journal of the American Water Resources Association, v. 39, no. 6, p. 1481–1496.

Borsuk, M.E., Stow, C.A., and Reckhow, K.H., 2002, Predicting the frequency of water quality standard violations—A probabilistic approach for TMDL development: Environmental Science and Technology, v. 36, no. 10, p. 2109–2115.

Breault, R.F., and Granato, G.E., 2003, A synopsis of technical issues for monitoring trace elements in highway and urban runoff, in Granato, G.E., Zenone, Chester, and Cazenas, P.A., eds., National highway runoff water-quality data and methodology synthesis, v. I—Technical issues for monitoring highway runoff and urban stormwater: Federal Highway Administration Report FHWA–EP–03–054, p. 165–233.

Bricker, O.P., 2003, An overview of the factors involved in evaluating the geochemical effects of highway runoff on the environment, in Granato, G.E., Zenone, Chester, and Cazenas, P.A., eds., National highway runoff water-quality data and methodology synthesis, v. I —Technical issues for monitoring highway runoff and urban stormwater: Federal Highway Administration Report FHWA–EP–03–054, p. 81–110.

Brouwer, Roy, and De Blois, Chris, 2008, Integrated modelling of risk and uncertainty underlying the cost and effectiveness of water quality measures: Environmental Modelling and Software, v. 23, p. 922–937.

Brown, J.W., ed., 2006, Eco-logical—An ecosystem approach to developing infrastructure projects: Federal Highway Administration Report FHWA–HEP–06–011, 96 p.

Buckler, D.R., and Granato, G.E., 2003, Assessing biological effects from highway-runoff constituents, in Granato, G.E., Zenone, Chester, and Cazenas, P.A., eds., National highway runoff water-quality data and methodology synthesis, v. I —Technical issues for monitoring highway runoff and urban stormwater: Federal Highway Administration Report FHWA–EP–03–054, p. 305–351.

Burton, G.A., Jr., and Pitt, R., 2002, Stormwater effects handbook—A tool box for watershed managers, scientists, and engineers: Boca Raton, Fla., CRC Press, 911 p.

Busby, M.W., 1966, Annual runoff in the conterminous United States, 1951–80: U.S. Geological Survey Hydrologic Investigations Atlas HA–212, 1 pl.

Butler, D., May, R.W.O., and Ackers, J.C., 1996, Sediment transport in sewers, Part 1—Background: Proceedings of the Institution of Civil Engineers, Water Maritime and Energy, v. 118, no. 2, p. 103–112.

California Department of Transportation, 2009, BMP pilot study guidance manual: California Department of Transportation, Division of Environmental Analysis, Stormwater Program, CTSW–RT–06–171.02.1, 368 p.

Canfield, D.E., Jr., and Bachmann, R.W., 1981, Prediction of total phosphorus concentrations, chlorophyll a and Secchi depths in natural and artificial lakes: Canadian Journal of Fisheries and Aquatic Sciences, v. 38, no. 4, p. 414–423.

Carmichael, G.R., 1982, Estimation of the drag coefficient of regularly shaped particles in slow flows from morphological descriptors: Industrial and Engineering Chemistry Process Design and Development, v. 21, p. 401–403.

Caruso, J.C., and Cliff, Norman, 1997, Empirical size, coverage, and power of confidence intervals for Spearman's rho: Educational and Psychological Measurement, v. 57, no. 4, p. 637–654.

Cazenas, P.A., Bank, F.G., and Shrouds, J.M., 1996, Guidance on developing water-quality action plans: Federal Highway Administration, Office of Natural Environment, 9 p.

Chalmers, A.T., Van Metre, P.C., and Callender, Edward, 2007, The chemical response of particle-associated contaminants in aquatic sediments to urbanization in New England, U.S.A.: Journal of Contaminant Hydrology, v. 91, p. 4–25.

Chen, Jieyun, and Adams, B.J., 2005, Analysis of storage facilities for urban stormwater quantity control: Advances in Water Resources, v. 28, p. 377–392.

Chen, Jieyun, and Adams, B.J., 2007, Development of analytical models for estimation of urban stormwater runoff: Journal of Hydrology, v. 336, p. 458–469.

Cheng, K.S., Ciang, J.L., and Hsu, C.W., 2007, Simulation of probability distributions commonly used in hydrological frequency analysis: Hydrological Processes, v. 51, p. 51–60.

Chow, V.T., 1954, The log-probability law and its engineering applications: Proceedings of the American Society of Civil Engineers, v. 80, no. 536, 25 p.

Chow, V.T., Maidment, D.R., and Mays, L.W., 1988, Applied hydrology: New York, McGraw-Hill, Inc., 572 p.

Church, P.E., Granato, G.E., and Owens, D.W., 2003, Basic requirements for collecting, documenting, and reporting precipitation and stormwater-flow measurements, in Granato, G.E., Zenone, Chester, and Cazenas, P.A., eds., National highway runoff water-quality data and methodology synthesis, v. I—Technical issues for monitoring highway runoff and urban stormwater: Federal Highway Administration Report FHWA–EP–03–054, p. 47–79.

Clary, Jane, Leisenring, Marc, and Hobson, Paul, 2011, International stormwater best management practices (BMP) database pollutant category summary—Metals: Water Environment Research Foundation, accessed January 1, 2012, at http://www.bmpdatabase.org/.

Clary, Jane, Leisenring, Marc, and Jeray, Joe, 2010, International stormwater best management practices (BMP) database pollutant category summary—Fecal indicator bacteria: Water Environment Research Foundation, accessed January 1, 2012, at http://www.bmpdatabase.org/.

Colorado State Department of Public Health and Environment, 2002, Colorado mixing zone implementation guidance: Denver, Colo., Colorado State Department of Public Health and Environment Water Quality Control Division, 64 p.

Conlon, K.J., and Journey, C.A., 2008, Evaluation of four structural best management practices for highway runoff in Beaufort and Colleton Counties, South Carolina, 2005–2006: U.S. Geological Survey Scientific Investigations Report 2008–5150, 121 p. (Also available at http://pubs.usgs.gov/sir/2008/5150/.)

Connecticut Department of Environmental Protection, 2010, Bathymetry and elevation: Connecticut Department of Environmental Protection, accessed January 1, 2012, at http://www.ct.gov/dep/cwp/view.asp?a=2698&q=322898&depNav_GID=1707&depNav=|#Bathymetry.

Cotton, A.P., and West, J.R., 1980, Field measurement of transverse diffusion in unidirectional flow in a wide, straight channel: Water Research, v. 14, no. 11, p. 1597–1604.

Day, T.J., 1977, Observed mixing lengths in mountain streams: Journal of Hydrology, v. 35, p. 125–136.

Devroye, Luc, 1986, Non-uniform random variate generation: New York, Springer-Verlag, 843 p.

Dietrich, W.E., 1982, Settling velocity of natural particles: Water Resources Research, v. 18, no. 6, p. 1615–1626.

Di Toro, D.M., 1984, Probability model of stream quality due to runoff: Journal of Environmental Engineering, v. 110, no. 3, p. 607–628.

Di Toro, D.M., Allen, H.E., Bergman, H.L., Meyer, J.S., Paquin, P.R., and Santore, R.C., 2001, Biotic ligand model of the acute toxicity of metals—1. Technical basis: Environmental Toxicology and Chemistry, v. 20, no. 10, p. 2383–2396.

Divine, C.E., Clemmer, R.L., Bilgin, Azra, Clonts, Jeff, and Giordano, T.J., 2007, Mixing zone characterization of two transition terrain streams with tracers: Journal of the American Water Resources Association, v. 43, no. 4, p. 864–874.

Dorman, M.E., Hartigan, J.P., Steg, R.F., and Quasebarth, T.F., 1996, Retention, detention, and overland flow for pollutant removal from highway stormwater runoff, v. I—Research report: Federal Highway Administration Report FHWA–RD–96/095, 179 p.

Dow, K.E., Steffler, P.M., and Zhu, D.Z., 2009, Case study—Intermediate field mixing for a bank discharge in a natural river: Journal of Hydraulic Engineering, v. 135, no. 1, p. 1–12.

Drexel University Math Forum, 1999, Deriving the Haversine formula: Philadelphia, Penn., Drexel University, accessed August 2, 2004, at http://mathforum.org/library/drmath/view/51879 html.

Driscoll, E.D., Di Toro, D.M., Gaboury, D., and Shelley, P.E., 1986, Methodology for analysis of detention basins for control of urban runoff quality: U.S. Environmental Protection Agency Report EPA440/5–87–001, 74 p.

Driscoll, E.D., Di Toro, D.M., and Thomann, R.V., 1979, A statistical method for assessment of urban runoff: U.S. Environmental Protection Agency Report EPA 440/3–79–023, 200 p.

Driscoll, E.D., Palhegyi, G.E., Strecker, E.W., and Shelley, P.E., 1989, Analysis of storm event characteristics for selected rainfall gages throughout the United States: U.S. Environmental Protection Agency OCLC 30534890, 43 p.

Driscoll, E.D., Shelley, P.E., Gaboury, D.R., and Salhotra, Atul, 1989, A probabilistic methodology for analyzing water quality effects of urban runoff on rivers and streams: U.S. Environmental Protection Agency Report EPA 841–R89–101, 128 p.

Driscoll, E.D., Shelley, P.E., and Strecker, E.W., 1990a, Pollutant loadings and impacts from highway stormwater runoff, v. I—Design procedure: Federal Highway Administration Report FHWA–RD–88–006, 67 p.

Driscoll, E.D., Shelley, P.E., and Strecker, E.W., 1990b, Pollutant loadings and impacts from highway stormwater runoff, v. III—Analytical investigation and research report: Federal Highway Administration Report FHWA–RD–88–008, 160 p.

Dupuis, T.V., 2002, Assessing the impacts of bridge deck runoff contaminants in receiving waters, v. 2, Practitioner's handbook: Washington, D.C., Transportation Research Board, National Cooperative Highway Research Program NCHRP Report 474, 93 p.

Edil, T.B., and Bosscher, P.J., 1994, Engineering properties of tire chips and soil mixtures: Geotechnical Testing Journal, v. 17, no. 4, p. 453–464.

Edwards, T.K., and Glysson, G.D., 1999, Field methods for measurement of fluvial sediment: U.S. Geological Survey Techniques of Water-Resources Investigations, book 3, chap. C2, 89 p.

Elshorbagy, A., Parasuraman, K., Putz, G., and Ormsbee, L., 2007, Deterministic and probabilistic approaches to the development of pH total maximum daily loads—A comparative analysis: Journal of Hydroinformatics, v. 9, no. 3, p. 203–213.

Emerson, C.H., and Traver, R.G., 2008, Multiyear and seasonal variation of infiltration from storm-water best management practices: Journal of Irrigation and Drainage Engineering, v. 134, no. 5, p. 598–605.

Falcone, J.A., Carlisle, D.M., Wolock, D.M., and Meador, M.R., 2010, GAGES–A stream gage database for evaluating natural and altered flow conditions in the conterminous United States: Ecological Archives E091–045–D1, accessed April 28, 2012, at http://esapubs.org/Archive/ecol/E091/045/default htm.

Fang, Xing, Cleveland, Theodore, Garcia, C.A., Thompson, David, and Malla, Ranjit, 2005, Literature review on timing parameters for hydrographs: Texas Department of Transportation Research Technical Report 0–4696–1, 83 p.

Fang, Xing, and Stefan, H.G., 1999, Projections of climate change effects on water temperature characteristics of small lakes in the contiguous U.S.: Climatic Change, v. 42, p. 377–412.

Federal Highway Administration, 1998, Environmental flow charts: Federal Highway Administration, Office of Natural Environment, 45 p.

Fischer, H.B., List, E.J., Koh, R.C.Y., Imberger, Jörg, and Brooks, N.H., 1979, Mixing in inland and coastal waters: New York, Academic Press, 483 p.

Fletcher, W.B., 2011, Oregon Department of Transportation stormwater management program—Research and monitoring: accessed January 1, 2012, at http://www.oregon.gov/ODOT/HWY/GEOENVIRONMENTAL/.

Fowler, G.D., 2008, Measuring suspended sediment characteristics to identify accurate monitoring techniques in stormwater runoff: Durham, N.H., University of New Hampshire, Department of Civil Engineering, Master of Science thesis, 194 p.

Frick, W.E., Roberts, P.J.W., Davis, L.R., Keyes, J., Baumgartner, D.J., and George, K.P., 2001, Dilution models for effluent discharges (4th ed.)—Visual Plumes: U.S. Environmental Protection Agency Report EPA/600/R-03/025, 137 p.

Gebert, W.A., Graczyk, D.J., and Krug, W.R., 1987, Average annual runoff map of the United States, 1951–80: U.S. Geological Survey Hydrologic Investigations Atlas HA–710, 1 pl.

Gentle, J.E., 2003, Random number generation and Monte Carlo methods (2d ed.): New York, Springer Science+Business Media, Inc., 381 p.

Geosyntec Consultants and Wright Water Engineers, 2009, Urban stormwater BMP performance monitoring, accessed January 1, 2012, at http://www.bmpdatabase.org/Docs/2009%20Stormwater%20BMP%20Monitoring%20Manual.pdf.

Gersib, Dick, 2011, Washington State Department of Transportation stormwater and watersheds program: Olympia, Wash., Washington State Department of Transportation, accessed January 1, 2012, at http://www.wsdot.wa.gov/Environment/waterquality/.

Gibbons, R.D., 2003, A statistical approach for performing water quality impairment assessments: Journal of the American Water Resources Association, v. 39, no. 4, p. 841–849.

Gibbs, R.J., Matthews, M.D., and Link, D.A., 1971, The relationship between sphere size and settling velocity: Journal of Sedimentary Petrology, v. 41, no. 1, p. 7–18.

Glysson, G.D., 1987, Sediment-transport curves: U.S. Geological Survey Open-File Report 87–218, 47 p.

Goforth, G.F., Heaney, J.P., and Huber, W.C., 1983, Comparison of basin modeling techniques: Journal of Environmental Engineering, v. 109, no. 5, p. 1082–1098.

Gottschalk, L.C., 1961, Reservoir sedimentation, Section 17, Part I, in Chow, V.T., ed., Handbook of applied hydrology: New York, McGraw-Hill Book Co., p. 17.1–17.33.

Granato, G.E., 1992, Theoretical investigation of particle fall velocity in an oscillating flow: Charlottesville, Va., University of Virginia, Department of Civil and Environmental Engineering, Master of Science thesis, 96 p.

Granato, G.E., 1999, Computer program for Point Location and Calculation of ERror (PLACER): U.S. Geological Survey Open File Report 99–99, 36 p.

Granato, G.E., 2006, Kendall-Theil Robust Line (KTRLine—version 1.0)—A Visual Basic program for calculating and graphing robust nonparametric estimates of linear-regression coefficients between two continuous variables: U.S. Geological Survey Techniques and Methods, book 4, chap. A7, 31 p., CD–ROM.

Granato, G.E., 2009, Computer programs for obtaining and analyzing daily mean streamflow data from the U.S. Geological Survey National Water Information System Web Site: U.S. Geological Survey Open-File Report 2008–1362, 5 appendixes, 123 p., CD–ROM.

Granato, G.E., 2010, Methods for development of planning-level estimates of stormflow at unmonitored sites in the conterminous United States: Federal Highway Administration Report FHWA–HEP–09–005, 90 p., CD–ROM.

Granato, G.E., 2012, Estimating basin lagtime and hydrograph-timing indexes used to characterize stormflows for runoff-quality analysis: U.S. Geological Survey Scientific Investigations Report 2012–5110, 47 p.

Granato, G.E., Bank, F.G., and Cazenas, P.A., 2003, Data quality objectives and criteria for basic information, acceptable uncertainty, and quality-assurance and quality-control documentation, in Granato, G.E., Zenone, Chester, and Cazenas, P.A., eds., National highway runoff water-quality data and methodology synthesis, v. I—Technical issues for monitoring highway runoff and urban stormwater: Federal Highway Administration Report FHWA–EP–03–054, p. 3–21.

Granato, G.E., Barlow, P.M., and Dickerman, D.C., 2003, Hydrogeology and simulated effects of groundwater withdrawals in the Big River Area, Rhode Island: U.S. Geological Survey Water Resources Investigations Report 03–4222, 76 p.

Granato, G.E., Carlson, C.S., and Sniderman, B.S., 2009, Methods for development of planning-level estimates of water quality at unmonitored stream sites in the conterminous United States: Federal Highway Administration Report FHWA–HEP–09–003, 53 p., CD–ROM.

Granato, G.E., and Cazenas, P.A., 2009, Highway-runoff database (HRDB version 1.0)—A data warehouse and preprocessor for the stochastic empirical loading and dilution model: Federal Highway Administration Report FHWA–HEP–09–004, 57 p.

Granato, G.E., and Smith, K.P., 1999, Estimating concentrations of road-salt constituents in highway-runoff from measurements of specific conductance: U.S. Geological Survey Water-Resources Investigations Report 99–4077, 22 p.

Gray, J.R., Glysson, G.D., Turcios, L.M., and Schwarz, G.E., 2000, Comparability of suspended-sediment concentration and total suspended solids data: U.S. Geological Survey Water-Resources Investigations Report 00–4191, 14 p.

Grossman, J.N., Grosz, A.E., Schweitzer, P.N., and Schruben, P.G., eds., 2008, The National Geochemical Survey—Database and documentation, v. 5.0: U.S. Geological Survey Open-File Report 2004–1001, accessed January 1, 2012, at http://tin.er.usgs.gov/geochem/doc/home htm.

Gualtieri, Carlo, and Mucherino, Carmela, 2008, Comments on "Development of an empirical equation for the transverse dispersion coefficient in natural streams" by Tae Myoung Jeon, Kyong Oh Baek, and Il Won Seo: Environmental Fluid Mechanics, v. 8, p. 97–100.

Guo, Qizhong, 2006, Correlation of total suspended solids (TSS) and suspended sediment concentration (SSC) test methods—Final report: New Jersey Department of Environmental Protection, Division of Science, Research, and Technology Report SR05–005, 52 p.

Guo, Qizhong, 2007, Effect of particle size on difference between TSS and SSC measurements—Restoring our natural habitat, in Proceedings of the 2007 World Environmental and Water Resources Congress, Tampa, Fla., May 15–19, 2007: Tampa, Fla., American Society of Engineers, 17 p.

Guo, Yiping, and Adams, B.J., 1998, Hydrologic analysis of urban catchments with event-based probabilistic models—v. I, Runoff volume: Water Resources Research, v. 34, no. 12, p. 3421–3431.

Gupta, M.K., Agnew, R.W., Gruber, D., and Kreutzberger, W.A., 1981, Constituents of highway runoff, in v. IV, Characteristics of runoff from operating highways—Research report: Federal Highway Administration Report FHWA/RD–81/045, 171 p.

Guthrie, R.C., and Stolgitis, J.A., 1977, Fisheries investigations and management in Rhode Island lakes and ponds: Rhode Island Division of Fish and Wildlife Fisheries Report no. 3, 256 p.

Guy, H.P., 1970, Fluvial sediment concepts: U.S. Geological Survey Techniques of Water-Resources Investigations, book 3, chap. C1, 55 p.

Guy, H.P., 1977, Laboratory theory and methods for sediment analysis: U.S. Geological Survey Techniques of Water-Resources Investigations, book 5, chap. C1, 64 p.

Haan, C.T., 1977, Statistical methods in hydrology: Ames, Iowa, Iowa State University Press, 378 p.

Hallermeier, R.J., 1981, Terminal settling velocity of commonly occurring sand grains: Sedimentology, v. 28, p. 859–865.

Harmel, R.D., Cooper, R.J., Slade, R.M., Haney, R.L., and Arnold, J.G., 2006, Cumulative uncertainty in measured streamflow and water quality data for small watersheds: Transactions of the American Society of Agricultural and Biological Engineers, v. 49, no. 3, p. 689–701.

Harrison, R.L., 2010, Introduction to Monte Carlo simulation, in Dubnickova, Anna, Dubnicka, Stanislav, Granja, Carlos, Leroy, Claude, and Stekl, Ivan, eds., Nuclear physics methods and accelerators in biology and medicine, Bratislava, Slovak Republic, July 6–15, 2009: Melville, New York, American Institute of Physics Conference, Proceedings 1204, p. 17–21.

Heard, S.B., Gienapp, C.B., Lemire, J.F., and Heard, K.S., 2001, Transverse mixing of transported material in simple and complex stream reaches: Hydrobiologia, v. 464, p. 207–218.

Hellekalek, P., 1998, Good random number generators are (not so) easy to find: Mathematics and Computers in Simulation, v. 46, p. 485–505.

Helsel, D.R., 2005, Nondetects and data analysis–Statistics for censored environmental data: Hoboken, N.J., John Wiley & Sons Publishing, Inc., 250 p.

Helsel, D.R., and Hirsch, R.M., 2002, Statistical methods in water resources—Hydrologic analysis and interpretation: U.S. Geological Survey Techniques of Water-Resources Investigations, book 4, chap. A3, 510 p.

Hem, J.D., 1992, Study and interpretation of the chemical characteristics of natural water (3d ed.): U.S. Geological Survey Water-Supply Paper 2254, 263 p.

Holnbeck, S.R., 2005, Sediment-transport investigations of the upper Yellowstone River, Montana, 1999 through 2001—Data collection, analysis, and simulation of sediment transport: U.S. Geological Survey Scientific Investigations Report 2005–5234, 69 p.

Horowitz, A.J., and Elrick, K.A., 1987, The relation of stream sediment surface area, grain size, and composition to trace element chemistry: Applied Geochemistry, v. 2, no. 4, p. 437–451.

Horowitz, A.J., Elrick, K.A., and Colberg, M.R., 1992, The effect of membrane filtration artifacts on dissolved trace element concentrations: Water Research, v. 26, no. 6, p. 753–763.

Horowitz, A.J., Elrick, K.A., and Smith, J.J., 2008, Monitoring urban impacts on suspended sediment, trace element, and nutrient fluxes within the City of Atlanta, Georgia, U.S.A.—Program design, methodological considerations, and initial results: Hydrological Processes, v. 22, p. 1473–1496.

Horwatich, J.A., Bannerman, R.T., and Pearson, Robert, 2011, Highway-runoff quality, and treatment efficiencies of a hydrodynamic-settling device and a stormwater-filtration device in Milwaukee, Wisconsin: U.S. Geological Survey Scientific Investigations Report 2010–5160, 75 p. (Also available at http://pubs.usgs.gov/sir/2010/5160/.)

Hosking, J.R.M., and Wallis, J.R., 1997, Regional frequency analysis—An approach based on L-moments: New York, Cambridge University Press, 224 p.

House, W.A., and Warwick, M.S., 1998, Hysteresis of the solute concentration discharge relationship in rivers during storms: Water Research, v. 32, no. 8, p. 2279–2290.

Huber, W.C., Cannon, LaMarr, and Stouder, Matt, 2006, BMP modeling concepts and simulation: U.S. Environmental Protection Agency EPA/600/R–06/033, 166 p.

Hughes, R.M., and Larsen, D.P., 1988, Ecoregions—An approach to surface water protection: Journal of the Water Pollution Control Federation, v. 60, no. 4, p. 486–493.

Interagency Advisory Committee on Water Data, 1982, Guidelines for determining flood-flow frequency: U.S. Geological Survey, Office of Water Data Coordination, Bulletin 17B of the Hydrology Subcommittee, 183 p. (Also available at http://water.usgs.gov/osw/bulletin17b/bulletin_17B.html.)

Intergovernmental Task Force on Monitoring Water Quality, 1995a, The strategy for improving water-quality monitoring in the United States: U.S. Geological Survey Final Report, 25 p.

Intergovernmental Task Force on Monitoring Water Quality, 1995b, The strategy for improving water-quality monitoring in the United States: U.S. Geological Survey Technical Appendixes, 117 p.

Jenerette, G.D., Lee, Jay, Waller, D.W., and Carlson, R.E., 2002, Multivariate analysis of the ecoregion delineation for aquatic systems: Environmental Management, v. 29, no. 1, p. 67–75.

Jeon, T.M., Baek, K.O., and Seo, I.W., 2007, Development of an empirical equation for the transverse dispersion coefficient in natural streams: Environmental Fluid Mechanics, v. 7, no. 4, p. 317–329.

Jirka, G.H., Doneker, R.H., and Hinton, S.W., 1996, User's Manual for CORMIX—A hydrodynamic mixing zone model and decision support system for pollutant discharges into surface waters: Office of Science and Technology, U.S. Environmental Protection Agency, 152 p.

Johnson, David, 1997, The triangular distribution as a proxy for the beta distribution in risk analysis: The Statistician, v. 46, no. 3, p. 387–398.

Kacker, R.N., and Lawrence, J.F., 2007, Trapezoidal and triangular distributions for Type B evaluation of standard uncertainty: Metrologia, v. 44, no. 2, p. 117–127.

Kent, K.M., 1973, A method for estimating volume and rate of runoff in small watersheds: U.S. Department of Agriculture, Soil Conservation Service, SCS–TP–149, 64 p.

Kim, J.Y., and Sansalone, J.J., 2008, Event-based size distributions of particulate matter transported during urban rainfall-runoff events: Water Research, v. 42, p. 2756–2768.

Kirby, W.H., 1972, Computer-oriented Wilson-Hilferty transformation that preserves the first three moments and the lower bound of the Pearson Type 3 distribution: Water Resources Research, v. 8, no. 5, p. 1251–1254.

Konstantinos, Fytianos, 2001, Speciation analysis of heavy metals in natural waters—A review: Journal of Association of Official Agricultural Chemists International, v. 84, no. 6, p. 1763–1769.

Kuzin, S.A., and Adams, B.J., 2010, Probabilistic approach to estimation of urban storm-water TMDLs—Regulated catchment: Journal of Irrigation and Drainage Engineering, v. 136, no. 9, p. 627–636.

Landers, M.N., Ankcorn, P.D., and McFadden, K.W., 2007, Watershed effects on streamflow quantity and quality in six watersheds of Gwinnett County, Georgia: U.S. Geological Survey Scientific Investigations Report 2007–5132, 62 p.

Langbein, W.B., and others, 1947, Topographic characteristics of drainage basins: U.S. Geological Survey Water-Supply Paper 968–C, p. 125–157.

Langseth, D.E., and Brown, Natalie, 2011, Risk-based margins of safety for phosphorus TMDLs in lakes: Journal of Water Resources Planning and Management, v. 137, no. 3, p. 276–283.

Leisenring, Marc, Clary, Jane, Lawler, Ken, and Hobson, Paul, 2011, International stormwater best management practices (BMP) database pollutant category summary—Solids: Water Environment Research Foundation, accessed January 1, 2012, at http://www.bmpdatabase.org/.

Leisenring, Marc, Clary, Jane, Stephenson, Julie, and Hobson, Paul, 2010, International stormwater best management practices (BMP) database pollutant category summary—Nutrients: Water Environment Research Foundation, accessed January 1, 2012, at http://www.bmpdatabase.org/.

L'Ecuyer, Pierre, 1998, Random number generation, chap. 4 *in* Banks, Jerry, ed., The handbook on simulation: New York, John Wiley, Inc., 66 p.

L'Ecuyer, Pierre, 1999, Good parameters and implementations for combined multiple recursive random number generators: Operations Research, v. 47, no. 1, p. 159–164.

L'Ecuyer, Pierre, and Simard, Richard, 2007, TestU01—A C library for empirical testing of random number generators: Association for Computing Machinery, Transactions on Mathematical Software, v. 33, no. 4, article 22, 40 p.

Li, Yingxia, Lau, Joo-Hyon, Kang, Sim-Lin, Kayhanian, Masoud, and Stenstrom, M.K., 2008, Optimization of settling tank design to remove particles and metals: Journal of Environmental Engineering, v. 134, no. 11, p. 885–894.

Li, Yingxia, Lau, Sim-Lin, Kayhanian, Masoud, and Stenstrom, M.K., 2006, Dynamic characteristics of particle size distribution in highway runoff—Implications for settling tank design: Journal of Environmental Engineering, v. 132, no. 8, p. 852–861.

Lide, D.R., ed., 1997, CRC handbook of chemistry and physics (77th ed.): New York, Chemical Rubber Company Press, 2,608 p.

Lin, Hong, 2003, Granulometry, chemistry and physical interactions of non-colloidal particulate matter transported by urban storm water: Louisiana State University and Agricultural and Mechanical College, Department of Civil and Environmental Engineering, Ph.D. dissertation, 266 p.

Linsley, R.K., Jr., Kohler, M.A., and Paulhus, J.L.H., 1975, Hydrology for engineers (2d ed.): New York, McGraw-Hill Book Company, 482 p.

Lopes, T.J., and Dionne, S.G., 2003, A review of semivolatile and volatile organic compounds in highway runoff and urban stormwater, *in* Granato, G.E., Zenone, Chester, and Cazenas, P.A., eds., National highway runoff water-quality data and methodology synthesis, v. I —Technical issues for monitoring highway runoff and urban stormwater: Federal Highway Administration Report FHWA–EP–03–054, p. 235–303.

Mackey, P.C., Barlow, P.M., and Ries, K.G., III, 1998, Relations between discharge and wetted perimeter and other hydraulic-geometry characteristics at selected streamflow gaging stations in Massachusetts: U.S. Geological Survey Water-Resources Investigations Report 98–4094, 44 p.

Maestre, Alex, Pitt, R.E., and Williamson, Derek, 2004, Nonparametric statistical tests comparing first flush and composite samples from the National Stormwater Quality Database, in James, W., ed., Models and applications to urban water systems: Guelph, Ontario, CHI Publications, v. 12, p. 317–338.

Mahler, B.J., and Van Metre, P.C., 2003, A simplified approach for monitoring hydrophobic organic contaminants associated with suspended sediment—Methodology and applications: Archives of Environmental Contamination and Toxicology, v. 44, no. 3, p. 288–297. (Also available at http://tx.usgs.gov/coring/pubs/Arch%20Env%20LVSS.pdf.)

Mahler B.J., Van Metre, P.C., and Callender, Edward, 2006, Trends in metals in urban and reference lake sediments across the United States, 1975–2001: Environmental Toxicology and Chemistry, v. 25, no. 7, p. 1698–1709.

Manis, Rod, Schaffer, Evan, and Jorgensen, Robert, 1988, Unix relational database management—Application and development in the Unix environment: Englewood Cliffs, N.J., Prentice Hall, 476 p.

Mansell, Martin, and Rollet, Fabien, 2006, Water balance and the behaviour of different paving surfaces: Water and Environment Journal, v. 20, p. 7–10.

Marsaglia, George, and Tsang, W.W., 2002, Some difficult-to-pass tests of randomness: Journal of Statistical Software, v. 7, no. 3, p. 1–9.

Marsalek, J.H., 1991, Pollutant loads in urban stormwater—Review of methods for planning-level estimates: Water Resources Bulletin, v. 27, no. 2, p. 283–291.

Marsalek, J.H., and Ng, Y.F., 1989, Evaluation of pollution loadings from urban nonpoint sources—Methodology and applications: Journal of Great Lakes Research, v. 15, no. 3, p. 444–451.

Massachusetts Division of Fisheries and Wildlife, 2010, State pond maps: Massachusetts Division of Fisheries and Wildlife, accessed January 1, 2012, at http://www.mass.gov/dfwele/dfw/habitat/maps/ponds/pond_maps htm.

Mathey, S.B., ed., 1998, National Water Information System: U.S. Geological Survey Fact Sheet FS–027–98, 2 p.

McCorquodale, J.A., 2007, Storm-water jets and plumes in rivers and estuaries: Canadian Journal of Civil Engineering, v. 34, no. 6, p. 691–702.

McCullough, B.D., 2008, Microsoft Excel's "Not the Wichmann–Hill" random number generators: Computational Statistics and Data Analysis, v. 52, p. 4587–4593.

McGill, Robert, Tukey, J.W., and Larsen, W.A., 1978, Variations of box plots: The American Statistician, v. 32, no. 1, p. 12–16.

McGowen, Scott, 2011, Caltrans statewide stormwater program: California Department of Transportation, accessed January 1, 2012, at http://www.dot.ca.gov/hq/env/stormwater/.

McMahon, Gerard, Gregonis, S.M., Waltman, S.W., Omernik, J.M., Thorson, T.D., Freeouf, J.A., Rorick, A.H., and Keys, J.E., 2001, Developing a spatial framework of common ecological regions for the conterminous United States: Environmental Management, v. 28, no. 3, p. 293–316.

Mykytka, E.F., and Cheng, C.Y., 1994, Generating correlated random variates based on an analogy between correlation and force in Tew, J.D., Manivannan, S., Sadowski, D.A., and Seila, A.F., eds., Proceedings of the 1994 Winter Simulation Conference of the Association for Computing Machinery, December 11–14, 1994, Lake Buena Vista, Fla., p. 1413–1416.

National Cooperative Highway Research Program, 2006, Evaluation of best management practices for highway runoff control: Washington, D.C., Transportation Research Board NCHRP Report 565, 143 p.

National Research Council, 2008, Urban stormwater management in the United States: Washington, D.C., National Academy of Sciences Committee on Reducing Stormwater Discharge Contributions to Water Pollution, 610 p.

New Jersey Division of Fish and Wildlife, 2008, Lake survey maps: New Jersey Division of Fish and Wildlife, accessed January 1, 2012, at http://www.state nj.us/dep/fgw/lakemaps.htm.

Novotny, Vladimir, 2004, Simplified data based total maximum daily loads, or the world is log-normal: Journal of Environmental Engineering, v. 130, no. 6, p. 674–683.

O'Connor, D.J., 1976, The concentration of dissolved solids and river flow: Water Resources Research, v. 12, no. 2, p. 279–294.

Omernik, J.M., 1995, Ecoregions–A spatial framework for environmental management, chap. 5, in Davis, W.S., and Simon, T.P., eds., Biological assessment and criteria—Tools for water resource planning and decision making: Boca Raton, Fla., Lewis Publishers, p. 49–66.

Omernik, J.M., 2004, Perspectives on the nature and definition of ecological regions: Environmental Management, accessed July 28, 2005, at http://springerlink metapress.com/openurl.asp?genre=article&id=doi:10.1007/s00267-003-5197-2.

Omernik, J.M., Chapman, S.S., Lillie, R.A., and Dumke, R.T., 2000, Ecoregions of Wisconsin: Transactions of the Wisconsin Academy of Sciences, Arts, and Letters, v. 88, p. 77–103.

Oregon State Department of Environmental Quality, 2007, Regulatory mixing zone internal-management directive part 2–Reviewing mixing zone studies: Oregon State Department of Environmental Quality Report 07–WQ–013–SWM–RN–00479 Rev. 1.1, 57 p.

Park, Daeryong, Loftis, J.C., and Roesner, L.A., 2011, Modeling performance of storm water best management practices with uncertainty analysis: Journal of Hydrologic Engineering, v. 16, no. 4, p. 332–344.

Patterson, C.C., and Settle, D.M., 1976, Accuracy in trace analysis, sampling, sample handling, and analysis: Washington, D.C., NBS Special Publication 422, National Bureau of Standards, 321 p.

Perry, J.H., ed., 1963, Chemical engineers' handbook: New York, McGraw-Hill Book Co., 2,522 p.

Pitt, Robert, Maestre, Alex, and Morquecho, Renee, 2008, National stormwater quality database (NSQD), version 3: Tuscaloosa, Ala., University of Alabama, Department of Civil and Environmental Engineering. (Also available at http://rpitt.eng.ua.edu/Research/ms4/mainms4.shtml.)

Pitt, Robert, and Vorhees, John, 2003, The design, use, and evaluation of wet detention ponds for stormwater quality management using WinDETPOND: Tuscaloosa, Ala., University of Alabama, Department of Civil and Environmental Engineering, 233 p. (Also available at http://rpitt.eng.ua.edu/Publications/StormwaterTreatability/Detention%20ponds%20with%20WinDETPOND.pdf.)

Poresky, Aaron, Clary, Jane, Strecker, Eric, and Earles, Andrew, 2011, International stormwater best management practices (BMP) database technical summary—Volume reduction: Water Environment Research Foundation, accessed January 1, 2012, at http://www.bmpdatabase.org/.

Press, W.H., Flannery, B.P., Teukolsky S.A., and Vetterling, W.T., 1992, Numerical recipes in Fortran 77—The art of scientific computing (2d ed.): New York, Cambridge University Press, 992 p.

Ramier, D., Berthier, E., Dangla, P., and Andrieu, H., 2006, Study of the water budget of streets—Experimentation and modeling: Water Science and Technology, v. 54, no. 6–7, p. 41–48.

Randall, A.D., 1998, Mean annual runoff, precipitation, and evapotranspiration in the glaciated northeastern United States, 1951–80: U.S. Geological Survey Open-File Report 96–395, 2 pls. (Also available at: http://ny.water.usgs.gov/pubs/of/of96395/OF96-395 html.)

Rao, R.A., and Delleur, J.W., 1974, Instantaneous unit hydrographs, peak discharges, and time lags in urban basins: Hydrological Sciences Journal, v. 19, no. 2, p. 185–198.

Reckhow, K.H., 1979, Quantitative techniques for the assessment of lake quality: U.S. Environmental Protection Agency Report EPA–440/5–79–015, 146 p.

Rhode Island Department of Environmental Management, 2010, Freshwater lake and pond water quality: Rhode Island Department of Environmental Management, accessed January 1, 2012, at http://www.dem ri.gov/programs/benviron/water/quality/surfwq/lakewq htm.

Rice, K.C., 1999, Trace-element concentrations in streambed sediment across the conterminous United States: Environmental Science and Technology, v. 33, no. 15, p. 2499–2504. (Data available at http://water.usgs.gov/nawqa/trace/pubs/est_rice_99 html.)

Riggs, H.C., 1968, Some statistical tools in hydrology: U.S. Geological Survey Techniques of Water-Resources Investigations, book 4, chap. A1, 39 p.

Robertson, D.M., Saad, D.A., and Wieben, A.M., 2001, An alternative regionalization scheme for defining nutrient criteria for rivers and streams: U.S. Geological Survey Water-Resources Investigations Report 2001–4073, 57 p.

Rossman, L.A., 2010, Storm water management model user's manual—Version 5.0: U.S. Environmental Protection Agency Report EPA/600/R–05/040, 285 p.

Sansalone, J.J., and Buchberger, S.G., 1997, Partitioning and first flush of metals in urban roadway storm water: Journal of Environmental Engineering, v. 123, no. 2, p. 134–143.

Sansalone, J.J., and Tribouillard, T., 1999, Variation in characteristics of abraded roadway particles as a function of particle size—Implications for water quality and drainage: Journal of the Transportation Research Record, no. 1690, p. 153–163.

Santore, R.C., Di Toro, D.M., Paquin, P.R., Allen, H.E., and Meyer, J.S., 2001, A biotic ligand model of the acute toxicity of metals—2. Application to acute copper toxicity in freshwater fish and *Daphnia*: Environmental Toxicology and Chemistry, v. 20, no. 10, p. 2397–2402.

Saucier, Richard, 2000, Computer generation of statistical distributions: U.S. Army Research Laboratory Report ARL–TR–2168, 105 p.

Sauer, V.B., Thomas, W.O., Jr., Stricker, V.A., and Wilson, K.V., 1983, Flood characteristics of urban watersheds in the United States: U.S. Geological Survey Water-Supply Paper 2207, 63 p.

Scherer, W.T., Pomroy, T.A., and Fuller, D.N., 2003, The triangular density to approximate the normal density—Decision rules-of-thumb: Reliability Engineering and System Safety, v. 82, p. 331–341.

Schneider, L.E., and McCuen, R.H., 2006, Assessing the hydrologic performance of best management practices: Journal of Hydrologic Engineering, v. 11, no. 3, p. 278–281.

Schueler, T.R., 1987, Controlling urban runoff—A practical manual for planning and designing urban BMPs: Washington, D.C., Metropolitan Washington Council of Governments, Department of Environmental Programs, no. 87703, 275 p.

Schueler, T.R., 1996, Irreducible pollutant concentrations discharged from stormwater practices: Watershed Protection Techniques, v. 2, no. 2, p. 361–363.

Schwartz, S.S., and Naiman, D.Q., 1999, Bias and variance of planning-level estimates of pollutant loads: Water Resources Research, v. 35, no. 11, p. 3475–3487.

Selbig, W.R., and Bannerman, R.T., 2011, Characterizing the distribution of particles in urban stormwater using fixed-point sample collection methods: U.S. Geological Survey Open-File Report 2010–1052, 14 p.

Sevin, A.F., 1987, Guidance for preparing and processing environmental and section 4(f) documents: Washington, D.C., Department of Transportation, Federal Highway Administration, Technical Advisory T–6640.8A, 62 p., accessed August 13, 2012, at http://environment.fhwa.dot.gov/projdev/impta6640.asp.

Shacklette, H.T., and Boerngen, J.G., 1984, Element concentrations in soils and other surficial materials of the conterminous United States: U.S. Geological Survey Professional Paper 1270, 105 p.

Shafer, M.M., Overdier, J.T., Phillips, Hugh, Webb, David, Sullivan, J.R., and Armstrong, D.E., 1999, Trace metal levels and partitioning in Wisconsin rivers: Water, Air, and Soil Pollution, v. 110, no. 3–4, p. 273–311.

Shamir, E., Imam, B., Morin, E., Gupta, H.V., and Sorooshian, S., 2005, The role of hydrograph indices in parameter estimation of rainfall-runoff models: Hydrological Processes, v. 19, p. 2187–2207.

Shiller, A.M., and Boyle, E.A., 1987, Variability of dissolved trace metals in the Mississippi River: Geochimica et Cosmochimica Acta, v. 51, p. 3273–3277.

Shirazi, M.A., Boersma, Larry, Johnson, C.B., and Haggerty, P.K., 2001, Predicting physical and chemical water properties from relationships with watershed soil characteristics: Journal of Environmental Quality, v. 30, p. 112–120.

Shoemaker, Leslie, Lahlou, Mohammed, Doll, Amy, and Cazenas, Patricia, 2000, Stormwater best management practices in an ultra-urban setting—Selection and monitoring: Federal Highway Administration Report FHWA–EP–00–002, 287 p.

Shuster, W.D., Zhang, Y., Roy, A.H., Daniel, F.B., and Troyer, M., 2008, Characterizing storm hydrograph rise and fall dynamics with stream stage data: Journal of the American Water Resources Association, v. 44, no. 6, p. 1431–1440.

Sinnott, R.W., 1984, Virtues of the haversine: Sky and Telescope, v. 68, no. 2, p. 159.

Slattery, M.C., and Burt, T.P., 1997, Particle size characteristics of suspended sediment in hillslope runoff and streamflow: Earth Surface Processes and Landforms, v. 22, p. 705–719.

Sloto, R.A., and Crouse, M.Y., 1996, HYSEP—A computer program for streamflow hydrograph separation and analysis: U.S. Geological Survey Water-Resources Investigations Report 96–4040, 46 p.

Smith, E.P., Ye, Keying, Hughes, Chris, and Shabman, Leonard, 2001, Statistical assessment of violations of water quality standards under section 303(d) of the Clean Water Act: Environmental Science and Technology, v. 35, no. 3, p. 606–612.

Smith, K.P., and Granato, G.E., 2010, Quality of stormwater runoff discharged from Massachusetts highways, 2005–07: U.S. Geological Survey Scientific Investigations Report 2009–5269, 198 p., with CD–ROM. (Also available at http://pubs.usgs.gov/sir/2009/5269/.)

Smith, R.A., Alexander, R.B., and Lanfear, K.J., 1993, Stream water quality in the conterminous United States—Status and trends of selected indicators during the 1980s, in Paulson, R.W., Chase, E.B., Williams, J.S., and Moody, D.W., compilers, National water summary 1990–91—Hydrologic events and stream water quality: U.S. Geological Survey Water-Supply Paper 2400, p. 111–140.

Smith, R.A., Alexander, R.B., and Schwarz, G.E., 2003, Natural background concentrations of nutrients in streams and rivers of the conterminous United States: Environmental Science and Technology, v. 37, no. 4, p. 3039–3047.

Smith, S.M., 2006, National geochemical database—Version 1.41: U.S. Geological Survey Open-File Report 97–492. (Also available at http://pubs.usgs.gov/of/1997/ofr-97-0492/.)

Stedinger, J.R., Vogel, R.M., and Foufoula-Georgiou, Efi, 1993—Frequency analysis of extreme events, chap. 18 *in* Maidment, D.R., ed., Handbook of Hydrology: New York, McGraw-Hill, Inc., p. 18.1–18.66.

Strecker, Eric, Mayo, Lynn, Quigley, Marcus, and Howell, James, 2001, Guidance manual for monitoring highway runoff water quality: Federal Highway Administration Report FHWA–EP–01–022, 206 p.

Strecker, E.W., Quigley, M.M., Urbonas, B.R., Jones, J.E., and Clary, J.K., 2001, Determining urban storm water BMP effectiveness: Journal of Water Resources Planning and Management, v. 127, no. 3, p. 144–149.

Thomann, R.V., and Mueller, J.A., 1987, Principles of surface water quality modeling and control: New York, N.Y., Harper Collins Publishers, p. 644.

Tipping, E., 1994, WHAM—A chemical equilibrium model and computer code for waters, sediments, and soils incorporating a discrete site/electrostatic model of ion-binding by humic substances: Computers & Geosciences, v. 20, no. 6, p. 973–1023.

Tipping, E., Lofts, S., and Lawlor, A.J., 1998, Modeling the chemical speciation of trace metals in the surface waters of the Humber system: The Science of the Total Environment, v. 210–211, p. 63–77.

Turcios, L.M., Gray, J.R., and Ledford, A.L., 2010, Summary of U.S. Geological Survey on-line instantaneous fluvial sediment and ancillary data: U.S. Geological Survey, Office of Surface Water, accessed August 13, 2012, at http://water.usgs.gov/osw/sediment/.

University of Florida Lakewatch, 2010, Lakes with bathymetric maps: Gainesville, Fla., accessed August 13, 2012, at http://lakewatch.ifas.ufl.edu/MapList.htm.

Urbonas, B.R., and Roesner, L.A., 1993, Chapter 28–Hydrologic design for urban drainage control; *in* Maidment, D.R., ed., Handbook of Hydrology: New York, McGraw-Hill, Inc., p. 28.1–28.52.

U.S. Copyright Office, 2000, Copyright law of the United States of America and related laws contained in Title 17 of the United States Code: Washington, D.C., U.S. Copyright Office Circular 92, 237 p.

U.S. Environmental Protection Agency, 1983, Technical guidance manual for performing wasteload allocation—Book 4, Lakes, reservoirs and impoundments, Chapter 3 Eutrophication: U.S. Environmental Protection Agency Report EPA 440/4–84–019, 173 p.

U.S. Environmental Protection Agency, 1985a, Rates, constants, and kinetics formulations in surface water quality modeling (2d ed.): U.S. Environmental Protection Agency Report 600/3–85–040, 455 p.

U.S. Environmental Protection Agency, 1985b, Water quality assessment—A screening procedure for toxic and conventional pollutants in surface and ground water, part 1 [revised]: U.S. Environmental Protection Agency Report EPA/600/6–85/002a, 628 p.

U.S. Environmental Protection Agency, 1985c, Water quality assessment—A screening procedure for toxic and conventional pollutants in surface and ground water, part 2 [revised]: U.S. Environmental Protection Agency Report EPA/600/6–85/002b, 494 p.

U.S. Environmental Protection Agency, 1986a, Technical guidance manual for performing wasteload allocation—Book 4, Lakes, reservoirs and impoundments, Chapter 3 Toxic substances impact: U.S. Environmental Protection Agency Report EPA 440/4–87–002, 311 p.

U.S. Environmental Protection Agency, 1986b, Technical guidance manual for performing wasteload allocation—Book 6, Design conditions, Chapter 1 Stream design flow for steady-state modeling: U.S. Environmental Protection Agency Report EPA 440/4–86–014, 321 p.

U.S. Environmental Protection Agency, 1987, Processes, coefficients, and models for simulating toxic organics and heavy metals in surface waters: U.S. Environmental Protection Agency Report EPA/600/3–87–015, 303 p.

U.S. Environmental Protection Agency, 1991, Technical support document for water quality-based toxics control: U.S. Environmental Protection Agency Report EPA/505/2–90–001, 315 p.

U.S. Environmental Protection Agency, 1992, National pollutant discharge elimination system storm water sampling guidance document: U.S. Environmental Protection Agency Report EPA 883–B–92–001, 177 p.

U.S. Environmental Protection Agency, 1994, Guidance for the data quality objectives process, EPA QA/G–4, Final report: U.S. Environmental Protection Agency Report EPA/600/R–96/055, 72 p.

U.S. Environmental Protection Agency, 1996a, Guidance for data quality assessment—Practical methods for data analysis, EPA QA/G–9, Final report: U.S. Environmental Protection Agency Report EPA/600/R–96/084, 164 p.

U.S. Environmental Protection Agency, 1996b, Guidelines for ecological risk assessment: U.S. Environmental Protection Agency Report EPA/630/R–95/002F, 188 p.

U.S. Environmental Protection Agency, 1996c, The metals translator—Guidance for calculating a total recoverable permit limit from a dissolved criterion: U.S. Environmental Protection Agency Report EPA 823–B–96–007, 60 p.

U.S. Environmental Protection Agency, 2000, Nutrient criteria—Technical guidance manual—Lakes and reservoirs (1st ed.): U.S. Environmental Protection Agency Report EPA–822–B00–001, 232 p.

U.S. Environmental Protection Agency, 2001, Risk assessment guidance for Superfund, v. 3, part A, Process for conducting probabilistic risk assessment: U.S. Environmental Protection Agency Report EPA 540–R–02–002, 385 p.

U.S. Environmental Protection Agency, 2002a, National recommended water quality criteria—2002: U.S. Environmental Protection Agency Report EPA–822–R–02–047, 43 p.

U.S. Environmental Protection Agency, 2002b, The twenty needs report—How research can improve the TMDL program: U.S. Environmental Protection Agency Report EPA–841–B–02–002, 43 p.

U.S. Environmental Protection Agency, 2003, Level III ecoregions of the continental United States: U.S. Environmental Protection Agency, National Health and Environmental Effects Research Laboratory, 1 pl., accessed January 1, 2005, at http://www.epa.gov/wed/pages/ecoregions/.

U.S. Environmental Protection Agency, 2005a, National management measures to control nonpoint source pollution from urban areas: U.S. Environmental Protection Agency Report EPA 841–B–05–004, 518 p.

U.S. Environmental Protection Agency, 2005b, Welcome to STORET, EPA's largest computerized environmental data system: U.S. Environmental Protection Agency, accessed October 1, 2011, at http://www.epa.gov/STORET/index.html.

U.S. Environmental Protection Agency, 2007a, An approach for using load duration curves in the development of TMDLs: U.S. Environmental Protection Agency Report EPA 841–B–07–006, 74 p.

U.S. Environmental Protection Agency, 2007b, Total maximum daily loads with stormwater sources—A summary of 17 TMDLs: U.S. Environmental Protection Agency Report EPA 841–R–07–002, 57 p.

U.S. Environmental Protection Agency, 2010, Water quality analysis simulation program (WASP): U.S. Environmental Protection Agency, accessed January 1, 2012, at http://www.epa.gov/athens/wwqtsc/html/wasp.html.

U.S. Environmental Protection Agency, 2011, Exposure assessment models— Hydrological simulation program FORTRAN (HSPF), at http://www.epa.gov/ceampubl/swater/hspf/.

U.S. Geological Survey, 2001, Collection and use of total suspended solids data: U.S. Geological Survey Office of Water Quality and Office of Surface Water Technical Memorandum no. 2001.03, 3 p., available at http://water.usgs.gov/admin/memo/SW/sw01.03.html.

U.S. Geological Survey, 2002, NWISWeb: New site for the Nation's water data: U.S. Geological Survey Fact Sheet FS–128–02, 2 p.

U.S. Geological Survey, 2011, National Water Information System Web interface—USGS water data for the Nation: U.S. Geological Survey, accessed October 1, 2011, at http://waterdata.usgs.gov/nwis.

U.S. Geological Survey, 2012, Instantaneous data archive: U.S. Geological Survey, accessed January 1, 2012, at: http://ida.water.usgs.gov/ida/.

Van Buren, M.A., Watt, W.E., and Marsalek, Jiri, 1997, Applications of the log-normal and normal distributions to stormwater quality parameters: Water Research, v. 31, no. 1, p. 95–104.

Van Metre, P.C., and Mahler, B.J., 2005, Trends in hydrophobic organic contaminants in urban and reference lake sediments across the United States, 1970–2001: Environmental Science and Technology, v. 39, no. 15, p. 5567–5574.

Van Metre, P.C., Mahler, B.J., and Furlong, E.T., 2000, Urban sprawl leaves its PAH signature: Environmental Science and Technology, v. 34, no. 19, p. 4064–4070.

Vice, R.B., Guy, H.P., and Ferguson, G.E., 1969, Hydrologic effects of urban growth: U.S. Geological Survey Water-Supply Paper 1591–E, 41 p.

Vogel, R.M., and Kroll, C.N., 1989, Low-flow frequency analysis using probability-plot correlation coefficients: Journal of Water Resources Planning and Management, v. 115, no. 3, p. 338–357.

Vogel, R.M., Rudolph, B.E., and Hooper, R.P., 2005, Probabilistic behavior of water-quality loads: Journal of Environmental Engineering, v. 131, no. 7, p. 1081–1089.

Vogel, R.M., Tsai, Y., and Limbrunner, J.F., 1998, The regional persistence and variability of annual streamflow in the United States: Water Resources Research, v. 34, no. 12, p. 3445–3459.

Vollenweider, R.A., 1975, Input-output models with special reference to the phosphorus loading concept in limnology: Swiss Journal of Hydrology, v. 37, p. 53–83.

Wagner, C.R., Fitzgerald, S.A., Sherrell, R.D., Harned, D.A., Staub, E.L., Pointer, B.H., and Wehmeyer, L.L., 2011, Characterization of stormwater runoff from bridges in North Carolina and the effects of bridge deck runoff on receiving streams: U.S. Geological Survey Scientific Investigations Report 2011–5180, 95 p., 8 appendix tables.

Waldron, M.C., and Archfield, S.A., 2006, Factors affecting firm yield and the estimation of firm yield for selected streamflow-dominated drinking-water-supply reservoirs in Massachusetts: U.S. Geological Survey Scientific Investigations Report 2006–5044, 39 p.

Walker, W.W., Jr., 1987, Phosphorus removal by urban runoff detention basins: Lake and Reservoir Management, v. 3, p. 314–326, accessed November 18, 2011, at http://www.wwwalker.net/pdf/dbasins.pdf.

Walker, W.W., Jr., 2007, P8 urban catchment model user's manual: Concord, Mass., IEP, Inc., 48 p., accessed November 18, 2011, at http://www.wwwalker net/ p8/index html.

Wanielista, M.P., Hardin, Mike, Runnebaum, Nicole, and Cohen, Ron, 2010, Recharge and runoff from urban impervious surfaces, in 2010 Annual Florida Stormwater Association Conference, Fort Myers, Fla., June 2010: Annual Florida Stormwater Association, accessed January, 1, 2012, at http://www.stormwater.ucf.edu/ research/FSAJune2010meeting.pdf.

Wanielista, M.P., and Yousef, Y.A., 1993, Stormwater management: New York, John Wiley and Sons, 579 p.

Warn, A.E., and Brew, J.S., 1980, Mass balance: Water Research, v. 14, p. 1427–1434.

Washington State Department of Ecology, 2010, Water quality program permit writer's manual: Olympia, Wash., Washington State Department of Ecology Publication no. 92–109, 764 p., available at http://www.ecy.wa.gov/programs/eap/ pwspread.html.

Wichura, M.J., 1988, Algorithm AS241—The percentage points of the normal distribution: Applied Statistics, Journal of the Royal Statistical Society, series C, v. 37, no. 3, p. 477–484.

Wiles, T.J., and Sharp, J.M., Jr., 2008, The secondary permeability of impervious cover: Environmental and Engineering Geoscience, v. 14, no. 4, p. 251–265.

Windom, H.L., Byrd, J.T., Smith, R.G., Jr., and Huan, Feng, 1991, Inadequacy of NASQAN data for assessing metal trends in the Nation's rivers: Environmental Science and Technology, v. 25, no. 6, p. 1137–1142.

Winter, T.C., 1981, Uncertainties in estimating the water balance of lakes: Water Resources Bulletin, v. 17, no. 1, p. 82–115.

Wong, T.H.F., Fletcher, T.D., Duncan, H.P., and Jenkins, G.A., 2006, Modeling urban stormwater treatment—A unified approach: Ecological Engineering, v. 27, no. 1, p. 58–70.

Wright Water Engineers and Geosyntec Consultants, 2007, Frequently asked questions fact sheet for the International Stormwater BMP Database—Why does the International Stormwater BMP Database project omit percent removal as a measure of BMP performance?: Water Environment Research Foundation, accessed January 1, 2012, at http://www.bmpdatabase.org.

Ying, G., and Sansalone, J.J., 2008, Granulometric relationships for urban source area runoff as a function of hydrologic event classification and sedimentation: Water Air Soil Pollution, v. 193, p. 229–246.

Yorke, T.H., and Herb, W.J., 1978, Effects of urbanization on streamflow and sediment transport in the Rock Creek and Anacostia River Basins, Montgomery County, Maryland, 1962–74: U.S. Geological Survey Professional Paper 1003, 71 p.

Yorke, T.H., Stamer, J.K., and Pederson, G.L., 1985, Effects of low-level dams on the distribution of sediment, trace metals, and organic substances in the lower Schuylkill River Basin, Pennsylvania: U.S. Geological Survey Water-Supply Paper 2256–B, 53 p.

Young, G.K., Stein, Stuart, Cole, Pamela, Kammer, Traci, Graziano, Frank, and Bank, F.G., 1996, Evaluation and management of highway runoff water quality: Federal Highway Administration Report FHWA–PD–96–032, 480 p.

Zoppou, Christopher, 2001, Review of urban storm water models: Environmental Modelling and Software, v. 16, no. 3, p. 195–231.

www.ingramcontent.com/pod-product-compliance
Lightning Source LLC
Chambersburg PA
CBHW081459170526
45166CB00008B/2484